ESSENTIAL STUDY GUIDE TO BHS STAGE 1

with Self-Assessment and Exam Tips

JULIE BREGA

J. A. Allen
LONDON

DEDICATION

This book is dedicated to Holly, Josh and George.

ACKNOWLEDGEMENTS

I'd like to thank Ilona Loftus for reading through and contributing to the content of this book, Erica Dorling for help with Section 9, Frances White for being the rider model, Diane Harvey who provided the basis of the text for Sections 11 and 12, and Carole Vincer and Dianne Breeze for the line drawings.

I'd also like to thank The British Horse Society for the assistance given in relation to the content of this book.

Finally, thanks to my editor Martin Diggle, and to Lesley Gowers of J. A. Allen for being very patient!

First published in Great Britain in 2011

ISBN 978 0 85131 979 7

J.A. Allen
Clerkenwell House
Clerkenwell Green
London EC1R 0HT

J.A. Allen is an imprint of Robert Hale Limited

www.allenbooks.co.uk

British Library Cataloguing in Publication Data
A catalogue record for this book is available from the British Library

Design by Judy Linard

Printed in Singapore by Craft Print International Limited

Contents

Acknowledgements 2
Introduction 5

1 Safety and Grooming 7
Safe Working Practice 8
Efficiency and Time Management 12
Basic Procedures in the Stable 13
Grooming 15
Notes on Foot Condition and Shoeing 23

2 Tail Bandages and Rugs 25
Tail Bandaging 25
Rugs 27

3 Tack 33
The Bridle 33
The Saddle 38
Handling Saddles and Bridles 42
Other Items of Tack 43
Tacking Up 46
Untacking 49
Tack Cleaning 50
Safety Checks 53

4 Horse Husbandry – Bedding and Mucking Out 57
Bedding 58
Mucking Out 61

5 Horse Husbandry – Describing Horses, Leading, Filling Haynets 67
Describing a Horse 67
Leading and Standing Up 73
Filling and Hanging Haynets 77

6 Clothing, Health, Fire Precautions, Accident Procedure, Riding Out, The BHS 79
Clothes for Working with Horses 79
Physical Well-being 81
Fire Precautions 81
Basic Accident Procedure 83
Riding on Public and Private Land 88
The British Horse Society 95

7 HORSE HEALTH AND BEHAVIOUR 97
Daily Inspections 97
Signs of Good Health 99
Signs of Ill Health 102
Horse Behaviour 111
Handling Horses 116

8 GRAZING 121
Safety at Grass 121
The Field 122
Care of the Paddock 127
Checking Horses at Grass 130

9 FEEDING AND WATERING 135
Feedstuffs 135
The Rules of Feeding 143

10 LEADING, MOUNTING AND ADJUSTING TACK 155
Preparation for Leading 156
Tack-checking Prior to Mounting 157
Mounting and Dismounting 159

11 SEAT AND BALANCE 163
The Correct Riding Position 163
Preparation for Jumping 167

12 AIDS, GAITS AND RIDING IN COMPANY 169
The Aids 170
Transitions, Straight Lines and Turns 174
The Gaits 177
Riding in Company 182

ITQ Answers 185

Introduction

People become involved with horses for many different reasons. Some ride for exercise and the challenge of acquiring new skills, to socialise and enjoy the countryside. Some simply want to look after horses for love of the horse.

There are many riders who attend their local riding school once a week, some of whom hope one day to be a horse owner. Whether you own a horse or not, so much can be gained from knowing more about the horse and his care.

This book has been written for the novice enthusiast, professional or not, with the particular aim of working towards the British Horse Society (BHS) Stage 1 Certificate in Horse Knowledge, Care and Riding.

Once you decide to take the BHS exam it is important that you check the British Horse Society website (www.bhs.org.uk) for exam updates – the syllabus will be updated from time to time. (This book is arranged in the order of the Spring 2011 BHS Syllabus.)

HOW TO USE THIS BOOK

The start of each section describes the required skills and knowledge covered in that section. You will also find a subject box, organised as follows:

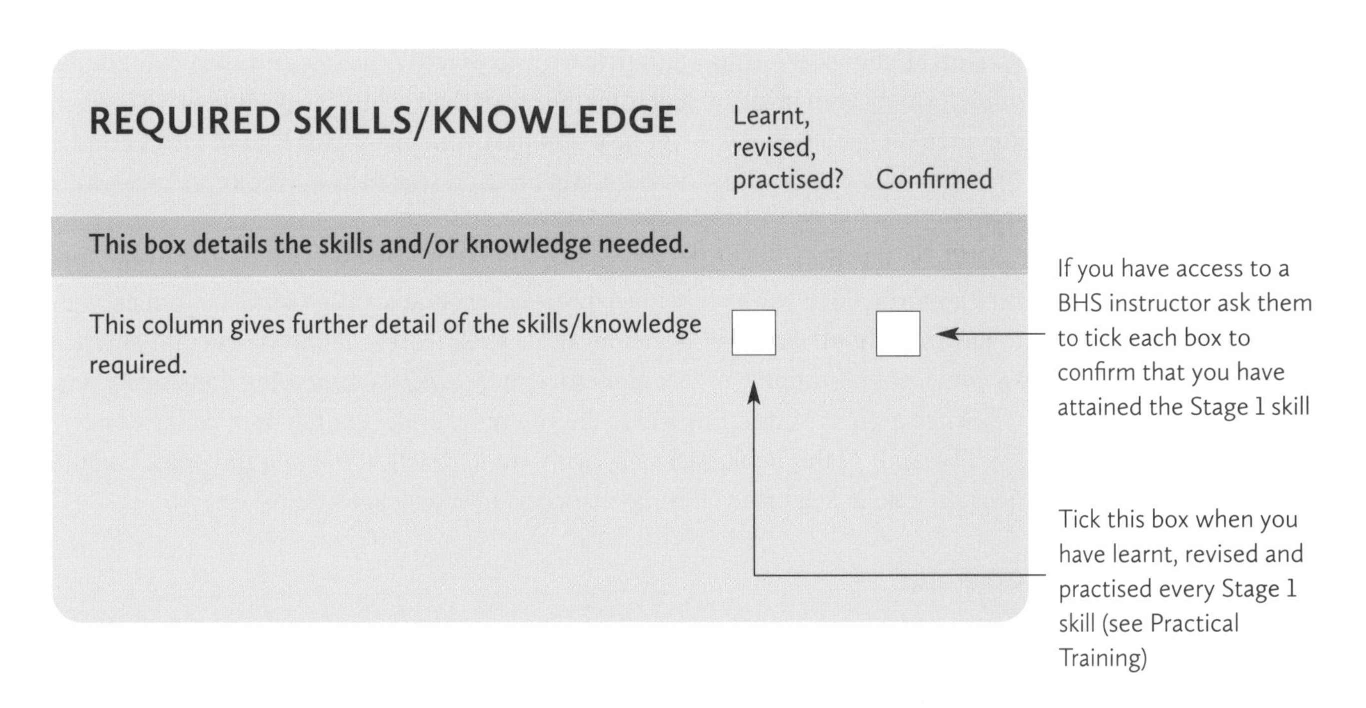

ITQ 0.0

What is an in-text question?

Throughout the book you will see in-text questions (ITQs) in boxes. These are revision aids for you – they are deliberately positioned somewhat further on than the relevant information in the main text. Write your answers in the box – you can check your answers at the end of the book.

EXAM TIP

Throughout the book you will find tips to help you in your exam.

SAFETY TIPS

Specific safety points are flagged up.

PRACTICAL TRAINING

As with all things, it is important to learn and practise the safe and correct methods of performing each practical skill from the outset; bad habits can be difficult to correct, are often dangerous and will not help you pass your exam. It is a good idea to find a BHS approved riding school or training centre and seek help with your practical skills training.

Whilst this book will help you prepare theoretically for your Stage 1 exam, it is very important that you gain as much practical experience as possible. It is not enough to just read about each topic – you must put the theoretical knowledge into practice so you become competent, safe and efficient. As well as improving your safety and efficiency around horses, this will make you much more confident in your exam.

The study of this book, backed up with sound practical training and practising the necessary skills will help you achieve success in your Stage exam.

Good luck!

1 Safety and Grooming

REQUIRED SKILLS/KNOWLEDGE	Learnt, revised, practised?	Confirmed
Work safely and efficiently.		
• Know how to use safe and efficient handling and working procedures, maintaining health, safety and welfare of yourself, others and horses at all times.	☐	☐
• Employ safe lifting and/or carrying techniques and be aware of potential hazardous situations.	☐	☐
• Maintain a clean working environment for yourself, others, horses and equipment.	☐	☐
Put on a headcollar and tie up the horse correctly.		
• Know how to approach a horse safely and correctly.	☐	☐
• Put on and adjust a headcollar correctly.	☐	☐
• Tie up the horse safely.	☐	☐
• Handle the horse safely in the stable.	☐	☐
Brush off/quarter a horse.		
• Select appropriate grooming kit items.	☐	☐
• Pick out the horse's feet and comment on the condition of the shoes.	☐	☐
• Brush off/quarter the horse effectively.	☐	☐
Be able to groom a horse.		
• Explain the use of items within a grooming kit.	☐	☐
• Understand the reasons for grooming.	☐	☐

SAFE WORKING PRACTICE

Handler training

- Many accidents are caused through lack of skill and experience on the part of the handler. Anyone working with horses, or even simply handling them, should receive thorough training in all aspects of their management. This should involve safety and accident prevention measures.
- Inexperienced handlers should work under close supervision and under the guidance of an experienced person. The inexperienced handler must not work with difficult or very young horses.
- When learning and training to work with horses you must aim for the highest standards in all areas, whether this be the actual care and management of the horses, the tidiness and organisation of the yard, or your general manner and style of communication with others. Attention to detail is vital when looking after horses.
- It is important that you are always aware of how the horse is reacting to you and to what is going on around him. Being 'in tune' with the horses will help you spot abnormalities promptly and also help prevent accidents.
- Horses are very labour-intensive animals to care for – you must learn to work safely and efficiently with due regard for the time taken to complete tasks, i.e. you must be safe and effective without wasting time.
- Horses should always be handled calmly, positively and confidently.

Working with horses is physically demanding and a level of general fitness and strength is necessary. Your fitness will develop the more you work with horses but the risk of injury increases for those who are not particularly fit or who suffer from health problems such as a bad back or arthritis.

Manual handling

When working on the yard, many heavy items have to be moved around, including bags of horse food, bales of hay and straw, plastic-wrapped bales of haylage, wheelbarrows full of muck and full water buckets.

Training should be given in correct handling procedures.

Moving heavy loads

- Estimate the weight of the load and if necessary seek help.
- Heavy items should be transported on a sack or wheelbarrow rather than being carried.
- When moving hay or straw bales wear gloves to protect your hands from the twine.
- To lift a bale or sack, start by standing close to it.

  - Square the sack up in front of you.
  - Keep your spine straight and bend your knees – don't lean over the sack to lift.
  - Grasp the sack securely and lift by straightening your knees rather than using arm strength alone.
- When carrying water buckets make sure the buckets are evenly filled and not too heavy. To even out the load and reduce the risk of back strain, carry a partially filled bucket in each hand, rather than one full bucket in one hand.
- When putting heavy items down, remember to bend your knees, not your back.
- If moving hay from a stack always take bales from the top – never pull out the lower bales, as the stack may collapse.

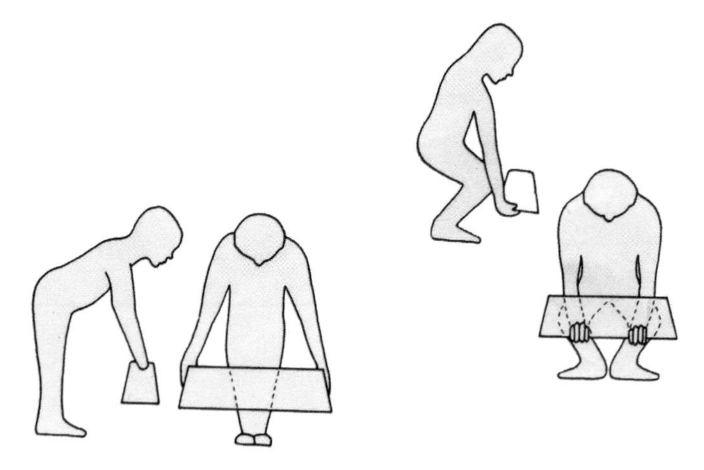

1.1 Correct lifting technique

General yard safety

- Keep the yard tidy at all times. It must be free of debris, tools, haynets, etc., which, apart from looking untidy, could injure a horse or trip someone over.
- Keep the yard gates closed at all times – then if a horse breaks loose, he is contained.
- To prevent horses from escaping, make sure stable door top and bottom bolts are securely fastened, especially last thing at night.
- Keep the feed room door closed. If an escapee horse gorges himself in the feed room he could get colic.
- Don't leave a horse loose in a stable wearing a nylon headcollar. The headcollar can become caught on the top bolt of the stable door, causing the horse to pull back and panic. In such a situation the nylon headcollar wouldn't break and the horse could be injured.
- Stables should be solidly constructed. Flimsy materials will splinter or break if the horse kicks out whilst rolling, etc. Make sure that stables are free of sharp projections such as nails.
- Stables should be well bedded down to prevent the horse from slipping. Rubber flooring is non-slip. Well banked-up sides can help to prevent the horse from

becoming cast (unable to rise, having lain down in an awkward position). Horses can damage themselves thrashing about when they become cast.

- Do not allow empty haynets to hang low in the stable as the horse may get a foot entangled.

ITQ 1.1

Why should the horse never be left wearing a nylon headcollar when he is loose in the stable?

IN-TEXT ACTIVITY

Write down one or two examples of accidents that have happened involving horses known to you.

With the benefit of hindsight, what steps could have been taken to prevent these accidents from occurring?

Personal protective equipment

When both handling and riding horses safety can be improved through the use of appropriate protective equipment.

Crash cap

The most important item is an approved safety standard crash cap that meets current PAS (Product Approval Specification) requirements. (You need to regularly check the latest safety standards as improvements to headgear are being made all the time.)

As from Spring 2011 the standards are:

PAS 015 with BSI Kitemark

BSEN 1384: 1997 with CE mark

ASTM F1163: 2004a with SEI mark

EN 1384: 1996 with CE mark

E2001 with Snell certification label

AS/NZ 3838: 2006 with SAI global mark

You should secure the chin strap before mounting. If borrowing a riding school hat give yourself time to select one that fits properly. Crash caps should be replaced if they sustain a heavy blow as a result of being dropped or involved in a fall.

Safe footwear

Many injuries are caused by horses jumping or simply standing on the handler's foot. Footwear must be water-resistant and stout. Flimsy shoes and sandals are not suitable when handling horses.

Ideally you should ride in jodhpur boots or long riding boots. These should have a smooth sole and a small yet defined heel. A wedged sole is dangerous as the foot can become stuck in the stirrup. Trainers also let the foot slip and are not robust enough to wear around horses. Wellingtons and walking boots tend to be broad and can trap the foot if they fit too snugly in the stirrup irons. Boots with buckles on the outside can get caught on the side of the stirrup.

Other protective items

Body protector. When jumping, especially for cross-country jumping, a current BSI standard body protector should be worn.

Gloves are recommended when riding, even in warm weather, as they prevent blisters caused by the rein pressure on the delicate skin between the ring and little fingers. Gloves specifically designed for riding will have reinforced material at this point, but driving gloves are often a suitable alternative. Gloves are also helpful when riding horses likely to get sweat on the reins, and should be worn when leading and lungeing horses.

Dust masks. When working in dusty conditions, e.g. when unloading hay or grooming, you may need to wear a dust mask, especially if you are prone to a dust allergy or asthma.

NB Jewellery, particularly dangly earrings and bracelets, should not be worn when handling horses as they can become caught, resulting in a ripped earlobe or broken wrist.

EFFICIENCY AND TIME MANAGEMENT

Without exception, professional involvement with horses is hard work. Depending on the type of yard, certain times of year may be quieter but on the whole horses are very labour-intensive animals to care for and most yards tend to be busy places.

The majority of the tasks need to be repeated every day, with many needing to be carried out more than once a day, e.g mucking/skipping out, feeding, watering, exercising, etc. and a great deal of time can be saved by planning the layout of the yard and organising the way in which you work to maximise efficiency.

Taking into account individual horses' temperaments you should aim to work to a high standard, quickly and safely. If you are hurried and sloppy you are more likely to make a mess, upset the horses and cause an accident.

Being set a good example at the start of your training and actually working with horses is the best way to learn good practice. Most employers very much value a yard assistant who works hard and maintains high standards.

Planning the layout of the yard and amenities, e.g. tack room, feed room, muck heap, etc. to improve ergonomics and keeping everything tidy and in its place saves time spent looking for and gathering tools and equipment, moving feed, hay, tack, wheelbarrows, etc.

ITQ 1.2 ?

Describe the safest way to pick up a full 20kg bag of horse food.

ITQ 1.3 ?

How can you help avoid back injury when moving heavy bales of hay?

ITQ 1.4

What is the safest way of carrying buckets of water across the yard?

ITQ 1.5

Describe how you will move a plastic wrapped bale of shavings from the barn to the stable yard.

BASIC PROCEDURES IN THE STABLE

Approaching the horse

Always speak when approaching the horse so he knows you are there, especially if he is not facing you. It is safest to approach at the shoulder. If you approach straight in front of the horse's face you may alarm him as some horses are slightly head-shy and if you approach at the horse's quarters you may get kicked.

Putting on a headcollar

Standing on the nearside, pass the lead-rope around the horse's neck. Slip his nose into the noseband of the headcollar and pass the headpiece behind his ears. Fasten the headcollar on the nearside. The noseband should sit two fingers' width down from the projecting cheekbone, allowing approximately one finger's width between the side of the horse's face and the noseband. Tuck all straps into their keepers.

1.2 Correctly fitting headcollar

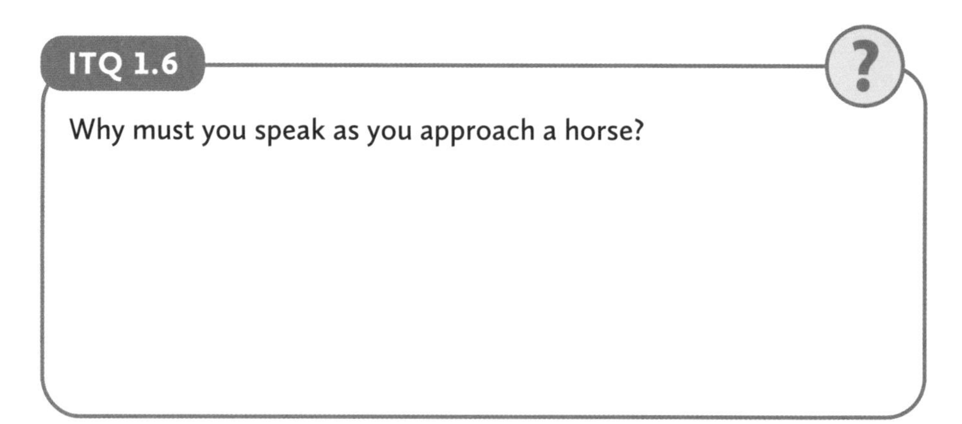

Leading through doorways and gateways

The safe technique for leading through doorways is one of the most important aspects to be taught when handling horses. Careless leading through doors and gates can cause serious injuries and unfortunately it is often seen.

The door or gate must be **fully open** and must **never** be allowed to swing onto the horse's hindquarters as he goes through. If the door does swing closed, the horse's hip may be caught, causing injury. He may then start to rush through doorways in an attempt to avoid being hurt again. You must walk through the doorway first to avoid being squashed between the horse and the door post.

Always lead the horse straight into the stable – do not try to lead him in on an angle as he may knock himself.

Tying up

As horses are so frequently tied up it is imperative that it is done safely. Tying a horse up can be dangerous if not carried out correctly. The tying ring must be:

- Securely attached to something solid – if it is fastened to an insecure object, should the horse pull back, it would come away and give him a nasty fright.
- Just at the horse's eye level.

A **weak link** must always be used – tie a piece of string in a loop through the tying-up ring and pass the lead-rope through this. The reason for this is that should the horse pull back and panic, the weak link will break rather than the headcollar or rope.

Always tie up the horse using a **quick-release knot**, to a solid object such as a tying ring fixed into the wall; never tie up to something insecure such as a rusty downpipe or loose piece of wood. The horse must never be tied directly to a gate or fence rail – should the horse pull back the rail or gate could come away which would panic the horse and cause serious injury.

Other points to note regarding tying up horses:

- Tie the horse up short enough to prevent him being able to put his foot over the rope, but not so short that he gets upset.

1.3 The quick-release knot

- Never leave a horse unattended when he is tied up.
- Whenever a horse is tied up in the yard, make sure that the yard gate is closed. The horse will then be contained in the event that he breaks loose. If tied in the stable the door must be closed for the same reason.
- Never tie up with the reins as this would hurt the horse's mouth should he pull back.
- Never tie horses close together as they may fight.

Horse tied correctly

Horse tied incorrectly

1.4 Tying up

GROOMING

Before you start to groom the horse you should skip the stable out (see Section 4) and remove the water bucket to prevent soiling with dust from the horse's coat. The haynet should also be removed and hung outside the stable safely.

Reasons for grooming

- Promotes cleanliness as dirt is removed.
- Promotes good health as waste products (e.g. sweat) are removed and the pores of the skin are kept clear.
- Prevents disease, saddle sores and girth galls.
- Improves the appearance of the horse.
- Allows the handler to 'get to know' the horse and check for cuts or swellings.
- Improves the circulation and muscle tone – this applies to both horse and groom!

ITQ 1.7

Make a note of four safety points to be observed when tying up horses:

1.

2.

3.

4.

ITQ 1.8

Give two examples of unsafe tying practices:

1.

2.

ITQ 1.9

Give two points to remember when leading a horse through a doorway:

1.

2.

1.5 The grooming kit

a. Dandy brush – used to remove dried mud and dirt.

b. Plastic curry comb – used to remove dried mud and dirt.

c. Rubber curry comb – used to remove loose hair.

d. Water brush – used to wash off stable stains and to 'lay' the mane.

e. Body brush – used to remove grease and dust; can also be used on the mane and tail.

f. Metal curry comb – used to clean the body brush.

g. Hoofpick – used twice daily (more frequently if necessary) to clean out the hooves. You must also have a rubber skip to pick the foot into.

h. Leather 'banger' or massage pad – used to strap the fit horse to tone the muscles.

i. Stable rubber – used to wipe off dust following strapping.

j Sponge – each horse should have his own separate sponges for eyes, nostrils and dock region, and a water bucket.

How to groom

Clothing/equipment notes:

Remove gloves.

Gather and check grooming kit, keeping it in a container.

Have to hand a small bucket partially filled with water.

Have to hand a headcollar and lead-rope.

Quartering

Quartering (brushing off) is a short grooming used to tidy up the horse or prepare him for a ride. After you have ridden it is correct to give your horse a more thorough grooming, known as **strapping**.

1. Put a headcollar on the horse and tie him up to prevent him from wandering around. If grooming in the stable, remove the water buckets to prevent them becoming soiled by the dust which is generated when grooming. If the horse's coat is particularly dusty and/or you have a dust allergy or suffer from asthma you should wear a protective dust mask.

2. Start by picking out the hooves. Position the skip behind the horse's foreleg. Stand facing the horse's tail and run your hand down the back of the foreleg and ask him to give his foot. You may need to take hold of some hair from the back of his heel – this hair is called **'feather'**. If he is reluctant to pick up his foot you may need to lean on him to make him shift his weight. This often makes it easier for you to pick up the foot. Hold the foot securely around the coronet/pastern.

   Use the **hoofpick** from heel to toe to prevent digging it into the frog or your own hand. Clean the foot out very thoroughly, making sure that you get all the dirt out from the clefts of the frog. If dirt is allowed to collect in the clefts it causes an ailment called **thrush**. (Thrush = a bacterial infection that can occur if there is a build-up of dirt and muck in the foot. The clefts of the frog soften and produce a foul-smelling black discharge.)

   Most horses will allow you to pick out all four feet from the nearside, which saves time.

   When holding the outside forefoot, bring it across behind the closer (nearside) foreleg.

   When holding the outside hind foot bring it across in front of the closer (nearside) hind leg.

   However, if the horse appears unfamiliar with this process, pick out the nearside hooves from the horse's nearside; the offside hooves from the horse's offside.

   Always clean the foot out into a **skip** to keep the yard or bedding clean. The hooves need to be picked out before and after a ride. Always clean out the feet at least once a day, preferably twice, even if you are not riding, and check on the condition of the shoes while you do so. The hooves should always be picked out when you bring the horse out of his stable and again upon return from exercise or the field.

SAFETY TIPS

- Position yourself safely, making sure you are not between the horse and the wall when picking out the hooves.
- Use the hoofpick from heel to toe so you do not accidently stick the hoofpick into your own arm.

As you pick out the feet in the Stage 1 exam you will be expected to comment on the condition of the horse's feet and shoes. You will mainly be looking to assess if the horse has been recently shod or if he looks as though he is in need of shoeing. These points are discussed in more detail in Notes on Foot Condition and Shoeing at the end of this section.

1.6 Picking out the foot

3. If the horse is rugged up on a cold day, undo the front buckle and belly straps of the rug. Knot the belly straps loosely together to stop them hanging down. Fold the front half of the rug back. Brush the exposed areas then replace the rug and refasten the front buckle. Fold back the rear half of the rug and brush off the hindquarters. Once this is done, fold the rug back down again, unknot and fasten the belly straps to prevent the horse from getting cold.

4. Your choice of brushes will depend on the thickness of the horse's coat, how sensitive the horse is likely to be and how dirty his coat is. If the horse has a reasonably thick coat and is not a 'ticklish', sensitive sort, use the **dandy brush** or **plastic curry comb** to remove the worst of the dried mud or dirt. Start at the top of the neck and brush briskly. When using the dandy brush you should use a flicking action at the end of each stroke to flick the dirt and dust off the coat. Fine-coated horses, e.g. Thoroughbreds, are likely to object to the use of a dandy brush or plastic curry comb – a **body brush** or soft dandy brush will be needed instead.

   Make sure you brush all over the horse; it is easy to miss areas such as the inside of the legs and underneath the belly. Take care when brushing around the flanks and belly as the horse is quite sensitive here. Careless brushing can make a horse bad-tempered and more likely to nip.

   If trying to remove a stable stain you can use the dandy brush backwards and forwards across the lie of the coat to help shift the stain. Once the stain has gone, brush the coat flat again. If preparing for a special occasion, e.g. a show, you could wash the stain off using the **water brush**. Always use a mild shampoo or soap and rinse it out well. Towel the wet areas as dry as possible – never wash a horse on a very cold day as there is a risk of him catching a chill.

5. Whilst working on the horse's neck, brush out his mane using the **body brush**. Starting just behind the poll, separate out a segment of hair and brush down to the roots to remove scurf. Continue down the length of his mane. Once you have brushed out the mane dampen the water brush and lay the mane down flat on the offside (right) of the neck. Dirt from the mane will have been flicked onto his neck so you should brush his neck after brushing out the mane.

1.7 Brushing the head

6. To groom the horse's head, untie the quick-release knot of the lead-rope and slip the headcollar around his neck. Steady the horse's head with your hand. Using the body brush, brush the head and forelock carefully, taking care not to bang the facial bones. Replace the headcollar and retie the quick-release knot.

SAFETY TIPS

- Never brush the horse's head without untying the quick-release knot as the horse may be head-shy and prone to pulling backwards.
- Never slip the headcollar around the horse's neck with the quick-release knot still tied. If the horse pulls back it will have a noose effect, causing the horse to panic.

7. To brush out the tail, stand to one side and hold the tail. Release a small amount of hair and brush out all knots using a body brush. Carry on doing this until the tail is free of tangles. You may find that using your fingers to untangle the hairs helps to prevent the hairs from breaking. A small amount of baby oil brushed through the tail prevents tangling. If the horse has a very thick tail you should use a fairly stiff body brush.

8. The last thing to do when quartering is to wipe the horse's eyes, nose and dock region with separate clean, damp **sponges**. Always wash the sponges well after use. Have a different coloured sponge for each area or use a permanent marker pen to label the sponges. To prevent the spread of disease, each horse should have his own sponges.

Strapping

Strapping is normally done after exercise because the coat is warm and loose and the pores of the skin are open. This means that the oil in the skin, which gives the coat its shine, will come out more easily – the dirt in the coat will also loosen off more easily. Any sweat on the coat will be brushed off when strapping, which promotes good hygiene.

1. Start off as with the quartering by tying up the horse and then picking out the hooves. Undo and fold back any rugs, as described earlier. Unless the horse is thin-skinned and sensitive, remove the worst of the dirt with the dandy brush.

2. Next, take the **body brush** in the hand nearer the horse's body and have the **metal curry comb** in the other hand. Starting at the top of the neck, use the brush vigorously in circular motions. Every three or four strokes, clean the brush on the metal curry comb. Occasionally tap the dirt out of the metal curry comb and you will see how much dirt you are getting out of his coat.

   The body brush has soft bristles and removes grease from the coat. This brush is not used on grass-kept horses as the grease in the coat acts as a waterproofer and helps to keep the horse warm.

   Groom the horse's entire body. Make sure you brush inside the horse's legs, around the girth area and under his belly

When grooming the neck on the side upon which the horse's mane lies, put the metal curry comb down and brush the mane out as previously described.

Groom the head as previously described.

3. If the horse is stable-kept and has a pulled tail, you can dampen the tail and put on a tail bandage to improve the appearance. However, never leave a tail bandage on for more than a few hours as it could interfere with circulation within the dock bone.

4. To tone up the horse's muscles you can 'bang' them with a wisp. This is referred to as **banging** or **wisping**. Use a folded **stable rubber** (a clean tea towel makes a good stable rubber, or alternatively a special leather **massage pad** or **banger** may be used), in a banging action on the main muscles on the body. This helps to tone the muscles and produce a shine. Most horses enjoy it once they get used to it. Wisping is done on the shoulders, neck and hindquarters – never wisp any bony areas, or over the loins. The kidneys lie beneath the loins, so wisping here would cause pain and damage.

5. Remove the last traces of dust by wiping the horse over with a stable rubber.

6. Sponge the eyes, nostrils and dock. Lay the mane and top of the tail with a damp water brush. Put hoof oil on the walls of the hooves for the final shine.

Some people advocate cleaning geldings' sheaths. Unless they are extremely dirty, sheaths should not be cleaned, particularly not with any soapy product. Male horses are supposed to have smegma, which contains a balance of healthy bacteria, in the sheath and excessive cleaning can cause irritation and disturb the natural balance of bacteria.

EXAM TIP

Practise grooming efficiently. The examiner will want to see you groom and handle the horse efficiently, in a workmanlike manner with due regard for safety at all times.

ITQ 1.10

What are the following used for when grooming?

a. Dandy brush

b. Body brush

c. Metal curry comb

d. Plastic curry comb

ITQ 1.11

What problem may occur if the hooves are not picked out regularly?

ITQ 1.12

What is the main difference between quartering and strapping?

ITQ 1.13

Why is strapping carried out after exercise?

NOTES ON FOOT CONDITION AND SHOEING

When to shoe

As a rough guide, horses should be re-shod every five to seven weeks. Even if the shoes do not appear unduly worn, it should not be left any longer than this because growth of the hooves will cause the feet to become unbalanced – and sometimes you will find that a horse needs shoeing more frequently than this. Ponies who don't go on the roads and are only in light work may be left unshod, but they will still need to have their hooves trimmed regularly as the hoof wall constantly grows downwards at a steady rate.

A horse needs shoeing when:

- Between five and seven weeks have elapsed since he was last shod.
- The shoe is loose or has slipped inwards.
- The clenches (nail tips) have risen from the hoof wall.
- A shoe has been cast (lost).
- The toe has grown over the shoe.
- The shoe has become excessively worn.
- The horn has cracked badly. This will reduce the security of the shoe, which may then be wrenched off.

Points to look for in a well-shod foot

To assess whether a horse has been well-shod, look for the following points:

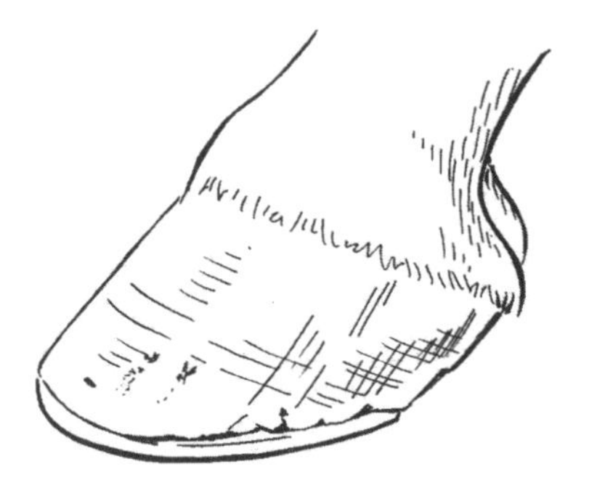
1.8 A foot in need of re-shoeing

1. The shoe must be made to fit the foot and not vice versa. If the toe has been cut back too much it is known as **dumping** and can adversely affect the balance of the foot.
2. The foot must be evenly reduced in size at the toe and heel, inside and outside of the foot.

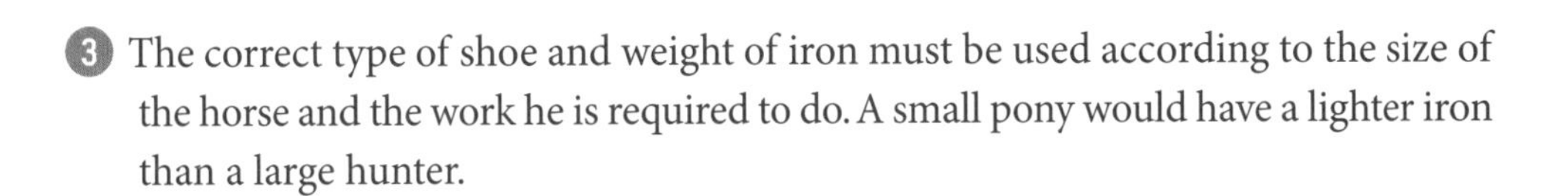

3. The correct type of shoe and weight of iron must be used according to the size of the horse and the work he is required to do. A small pony would have a lighter iron than a large hunter.

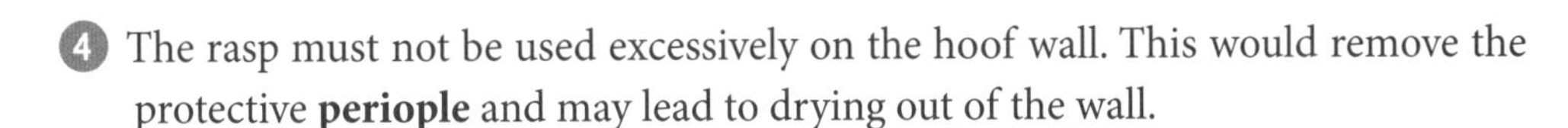

4. The rasp must not be used excessively on the hoof wall. This would remove the protective **periople** and may lead to drying out of the wall.
5. There should not be excessive use of the knife on the sole or frog. The frog should ideally be in contact with the ground – it cannot perform its functions if it does not touch the ground.

6. The correct number of nails has been used. There are normally seven, three on the inside and four on the outside. However, there may be occasions when it is necessary for the farrier to use a different number and/or space them differently when dealing with a problem with the horse's foot.

7. The correct size of nail must be used. The nail head should be neither too large nor too small.

8. The clenches must be even and approximately one-third of the way up the wall.

9. No daylight must show between the foot and the shoe.

10. The groove for the toes and quarter clips must be neat. The clips must be neatly embedded in.

11. The heels must not be too short or too long.

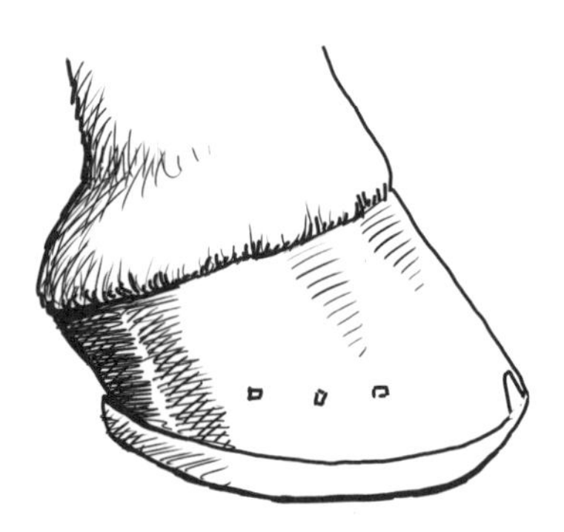

1.9 A well-shod foot

2 Tail Bandages and Rugs

REQUIRED SKILLS/KNOWLEDGE	Learnt, revised, practised?	Confirmed
Know how to put on and take off a tail bandage.		
• Put on a tail bandage.	☐	☐
• Remove a tail bandage.	☐	☐
Know how to put on and take off rugs.		
• Identify different rugs and their method of securing.	☐	☐
• Know how to rug up and remove rugs safely.	☐	☐

TAIL BANDAGING

The tail bandage is used to:

- Protect the dock bone from rubbing when travelling in a lorry or trailer.
- Improve the appearance of the tail.
- When a mare is being covered (mated) or about to foal she will wear a tail bandage.

Applying a tail bandage

No padding is required beneath a tail bandage. The horse must be tied up when you put on a tail bandage to prevent him from wandering around. Before bandaging, you may dampen the top of the tail to help the hairs lie flat. *Never use a wet bandage* as it may shrink and cut off the circulation in the dock area – for the same reason a tail bandage must not be left on overnight.

SAFETY TIP

- Show awareness of safety – make sure the horse knows you are there. Approach him at the shoulder and run your hands towards his hindquarters. Do not go straight to his tail or you may alarm him and cause him to kick.

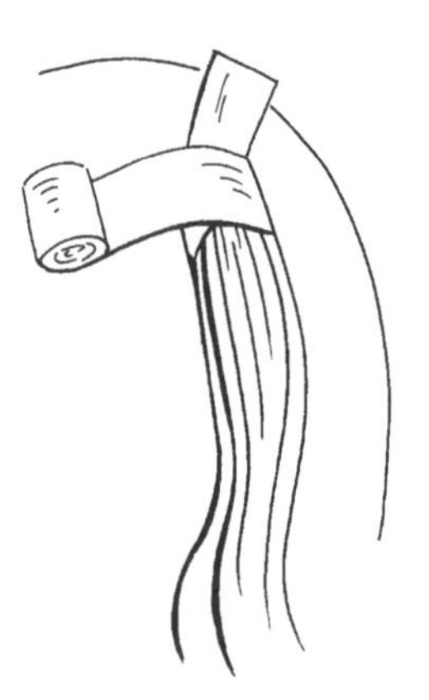

1. Standing behind the horse, unroll 15cm (6in) of bandage. Place the bandage under the tail and wrap securely around the top of the tail. This is the most important, but difficult, part. You must get the bandage securely started to prevent the whole thing from slipping down. Start to wind the bandage around the top of the tail.

2. Bandage down to the end of the dock bone and back up again until you run out of bandage, keeping it as neat as possible. Overlap each time by 50 per cent, keeping the bandage lines straight.

3. Keep the tension even.

4. Tie the strings on the back of the tail, ensuring they are no tighter than the bandage. Tie the strings neatly and fold a piece of the bandage down over the knot to cover it completely. This stops the strings from hanging down, which looks untidy, and reduces the chance of the bandage coming undone.

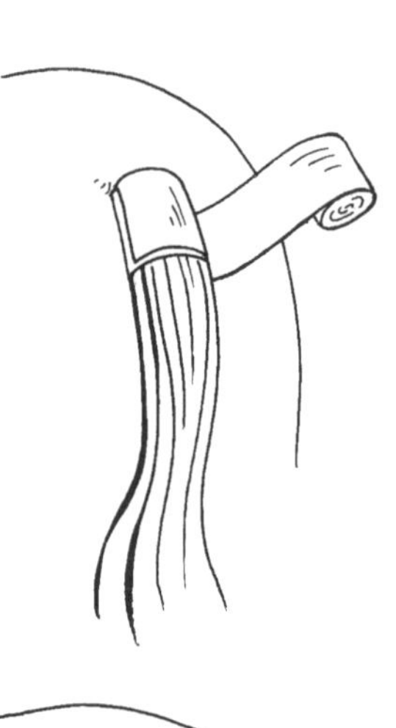

ITQ 2.1

Why must you never put a wet tail bandage on a horse's tail?

ITQ 2.2

Why are tail bandages applied?

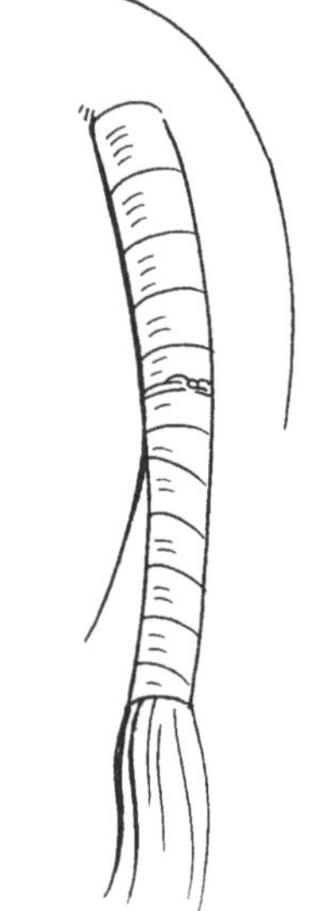

2.1 Applying a tail bandage

Removal

To remove the tail bandage, undo the ties and carefully pull downwards on the bandage. The skin on the underside of the dock is very soft so you must not pull roughly. Also beware that your fingernails don't scrape the underside of the dock.

If the tail is plaited you must unwind the bandage rather than pulling it down to prevent spoiling the tail plait.

Do not rewind the bandage as you remove it. If the bandage is soiled it should be washed and dried before being re-rolled.

EXAM TIP

Practise rolling tail bandages neatly and efficiently. Fold the tabs up and roll them into the bandage, pulling the bandage tight as you roll it.

RUGS

Stable rug (night rug)

The old-fashioned stable rugs were made from jute, a heavy woollen-lined hessian-type material. There is now a wide range of quilted synthetic rugs available in a variety of styles, weights and thicknesses. There are specially designed under-rugs for additional warmth.

One consequence of the changes in rug design is that the means of fastening them has changed. Traditional rugs were usually held in place by a **roller** or **surcingle**. A roller is a padded strap which buckles on the nearside. If a roller is used, extra padding such as a square of thick sponge should always be placed beneath it to stop the roller pressing on the spine. A roller can be used to hold extra blankets in position beneath the night rug.

A **surcingle** is similar to a roller but has no padding of its own, so it is essential that, when used, padding should be placed beneath it to protect the horse's spine. Surcingles are sometimes used to hold anti-sweat rugs or day rugs in position.

The majority of stable rugs nowadays are held in position by cross-over belly straps which pass beneath the horse's abdomen. Some have adjustable interlocking leg straps. The advantage of these straps over rollers and surcingles is that there is never any pressure on the horse's spine – as a result, rollers and surcingles are becoming much less common.

All rugs without leg straps must have a fillet string or strap to prevent the rug from slipping forwards.

Turnout (New Zealand) rugs

Turnout rugs (often called New Zealand rugs) are waterproof, designed for outdoor use, to be worn when the horse is in the field.

There are many types of turnout rug on the market. They vary in weight and thickness and most now have cross-over belly straps instead of interlocking hind leg straps, which were once popular. Some old-fashioned turnout rugs have surcingles attached, the main danger of which is that too much pressure will be exerted on the spine. Most of the new types of rug are made from synthetic non-rip material, making

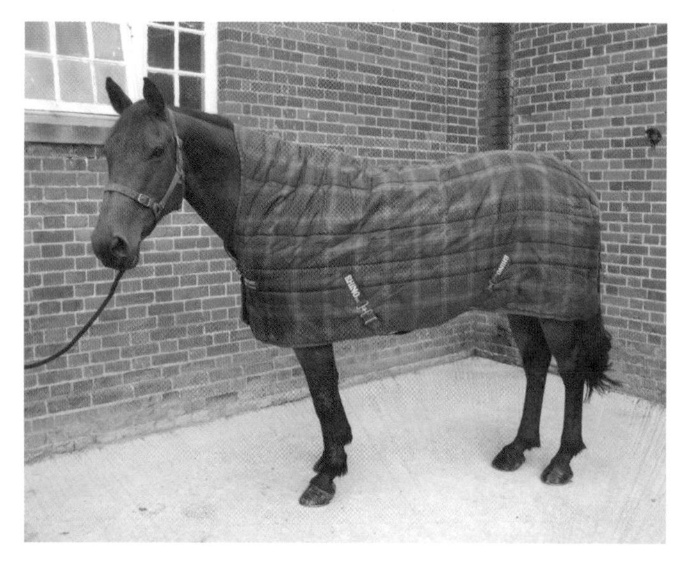

2.2 Stable rug

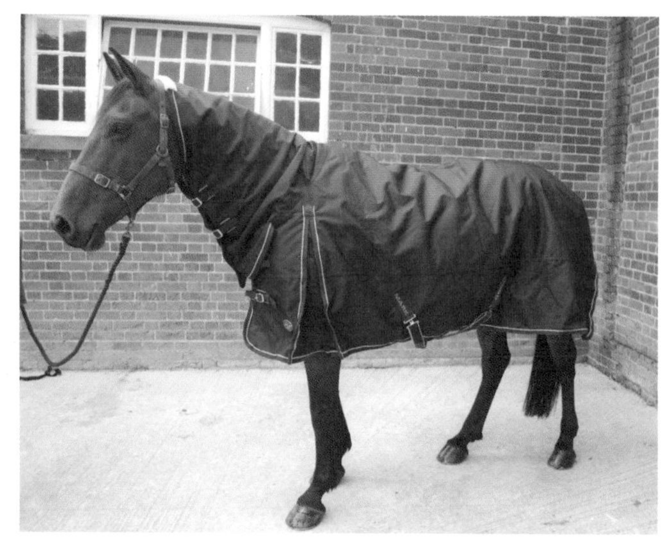

2.3 Turnout rug

them lighter than the old-fashioned canvas ones.

If a rug does have interlocking hind leg straps they should be fastened fairly loosely around the hind legs. If it does not have interlocking leg straps it must have a fillet strap which passes under the tail to stop the rug from slipping forwards or blowing up over the horse's back in windy weather.

You must always check that the rug does not slip too far back and chafe the horse's shoulders and that the leg straps do not rub the insides of the hind legs. It is a good idea to have two turnout rugs so that one can be drying off or cleaned whilst the horse wears the other. (In the UK climate, this can be considered essential.)

Other types of rugs

Anti-midge sheets

Lightweight anti-midge rugs are available for horses who suffer from sweet itch (an allergic reaction to the saliva of biting midges) to wear in the field.

Anti-sweat sheets

These are lightweight cotton mesh rugs used to prevent a horse who has sweated up from catching a chill whilst he dries off in the stable (or whilst being walked in hand after competing). They can be used beneath another rug to allow pockets of warm air to be trapped, helping to dry the horse. Once an anti-sweat sheet is wet it must be replaced with a dry one. The modern cooler rug has virtually superseded anti-sweat sheets.

Cooler/thermal rugs

These are specially designed to draw moisture away from the horse's coat and to promote cooling and drying without chilling. Most cooler rugs are fitted with cross-over belly straps and a fillet string.

Day rugs

These are normally used when travelling to a show and are smart woollen rugs with coloured binding, often with matching surcingles. They are not normally used in the stable, as they would be easily spoilt. Woollen day rugs can make a horse sweat and the sweat is not easily absorbed by the wool, so cooler/thermal rugs are often more appropriate.

IN-TEXT ACTIVITY

Obtain an equestrian supplies catalogue and find one example of each of the following:

	Rug name	Cost
Winter-weight turnout rug		
Lightweight shower-proof turnout sheet		
Winter-weight stable rug		
Lightweight cooler rug		

ITQ 2.3 ?

What is the main disadvantage of rollers and surcingles?

Rugging up and unrugging

Rugging up

1. The horse should be tied up to prevent him from wandering around.
2. Fold the rug in half so that the back half lies on top of the front half.
3. From the nearside, place the rug over the horse's withers and straighten. Unfold the back half. Do not pull the rug too far back – the end of the rug should just

reach the top of the tail. Rugs tend to slip back so put it on well forward so it is loose around the shoulders. A rug that is tight on the shoulders may cause chafing and soreness.

2.4 Rugging up

4. If the rug has belly straps, fasten the front buckles first, then the cross-over belly straps. These should not be too tight against the horse's belly. Likewise, they must not be so loose that they hang down in loops. The horse may put a foot through a loop whilst lying down in the stable.

There is, however, a disadvantage to having only the front buckles fastened. This is explained in point 5. opposite.

(If the rug has a surcingle, the order of buckle fastening is different. Put on the roller or surcingle – with thick sponge for padding – just behind the withers, with buckles on the nearside. Check the offside in case the roller is twisted. Fasten the

SAFETY TIP 

- The reason for fastening the front buckles first is to prevent the horse from standing with the rug held in place only by the belly straps as the rug may, potentially, slip backwards, causing the horse to panic as the belly straps would become entangled around his hind legs.

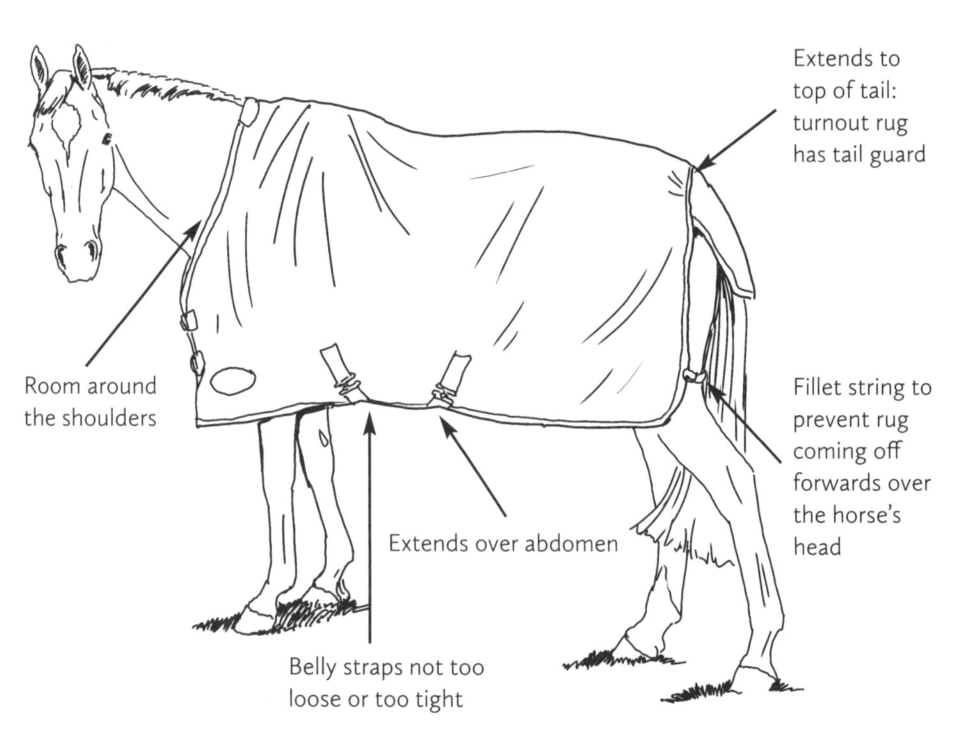

2.5 Correct rug fitting points

buckles on the nearside quite tightly and smooth out creases in the rug, beneath the roller.)

5. Next, fasten the front buckle. A rug that is fastened by the front buckles only is in danger of slipping around the horse's chest. If the rug has hind leg straps, for example on a New Zealand rug, you should always fasten these last.

   Always finish rugging up properly before leaving the horse. Don't leave him standing with only the front buckles or belly straps fastened for any length of time.

6. In cold weather it may be necessary to use more than one rug. Some rugs are fitted with neck hoods for additional warmth. If using a blanket beneath a jute rug, fold the blanket in half and place it well up onto the horse's neck. Unfold the lower half to cover the back, just reaching the top of the tail. Next, put on the top rug in the way described earlier. Fold the two front corners of the blanket to the withers, forming a 'V' shape. Fold the 'V' back down. It should cover the withers and part of the horse's backbone. Put the roller and padding on top of this 'V' shape and fasten to hold the blanket in position and prevent it from slipping back. Then fasten the front buckle.

   To keep the horse warm and to prevent pressure on the spine, it is more satisfactory to use additional stable rugs which have belly straps, rather than having to use a roller.

2.6 A correctly fitted underblanket

Removing a rug

1. Check to see if the rug has hind leg straps. If it has, undo them first. Once undone, refasten the clips to prevent the straps from hanging down. Next, undo the belly straps and then the front buckle. If the rug has a roller, undo the front straps first, finally, undo and remove the roller.

2. The rug can then be simply pulled off backwards over the horse's tail and folded up, or it can be folded, front half over back, and then removed.

3. The rug must then be hung up safely, not placed on the ground where it can get soiled and become a tripping hazard.

ITQ 2.4

How can you tell if a rug fits the horse correctly?

ITQ 2.5

Why should a horse not be left standing with only the belly straps holding a rug in place?

ITQ 2.6

Why should a horse not be left standing with only the front buckle of the rug fastened?

3 Tack

REQUIRED SKILLS/KNOWLEDGE	Learnt, revised, practised?	Confirmed
Put on and remove a saddle, bridle and martingale or hunting breastplate.		
• Put on a bridle, with a noseband, correctly.	☐	☐
• Know how to put on a saddle with a correctly fitted numnah or saddlecloth.	☐	☐
• Put on a martingale or hunting breastplate.	☐	☐
• Check tack for safety and comfort.	☐	☐
• Untack safely and efficiently.	☐	☐
Demonstrate knowledge about tack and rugs.		
• Identify parts of the bridle.	☐	☐
• Identify parts of the saddle.	☐	☐
• Discuss the quality of the tack used, recognising whether it is worn and/or damaged.	☐	☐
• Understand the consequences of using worn or dirty tack.	☐	☐
• Know how to clean tack correctly.	☐	☐

THE BRIDLE

Bridles may be bought in pony, cob or full sizes, with various sizes, colours and weights of leather used. The quality of the leather will determine the cost of the bridle. The bit is normally purchased separately.

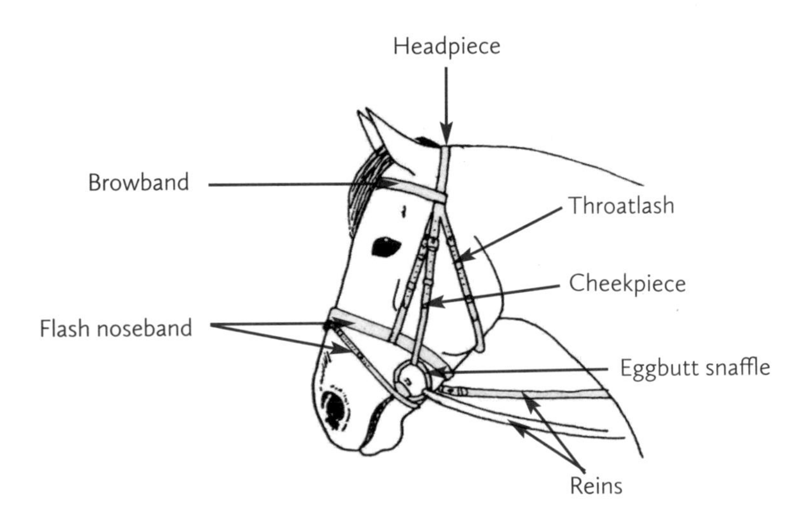

3.1 The snaffle bridle, shown with a Flash noseband

Main parts of the bridle

The **headpiece** passes behind the horse's ears. The **throatlash** extends down from the offside of the headpiece and buckles up on the nearside. It prevents the bridle from being pulled forwards over the horse's head, in the event of the rider taking a fall. To prevent interference with the horse's windpipe as he flexes his jaw when working, it must be buckled up to allow the width of a clenched fist between it and the horse's cheek.

There are two **cheekpieces.** One attaches on either side of the headpiece and together they hold the bit in place. They should be evenly adjusted on the same hole either side. Ideally, if the bridle is a good fit for the horse, this is halfway down the cheekpieces, which allows for adjustment up or down.

The **browband** attaches to the headpiece, preventing it from slipping back. It should not be too tight or it will pull the headpiece forward and pinch the ears.

The **bit** must not protrude more than 1cm (about ½in) either side of the mouth and must be just high enough to create approximately two wrinkles in the corners of the lips.

Types of noseband

There are many different types of **noseband** available.

Cavesson noseband

This is often worn to make the bridle look complete. It has no effect on the horse's jaw but can be used for attaching a standing martingale. It should be fitted to allow two fingers' width from the projecting cheekbones and be loose enough to permit two fingers' width between the front of the horse's nose and the noseband.

Drop noseband

This is fitted one hand's width from the upper edge of the horse's nostril, on the bony part of the face, and should be buckled around and under the bit. It should be tight enough to keep the horse's jaws closed but not *so* tight that it pinches the skin. The drop is designed to prevent the horse from opening his mouth and evading the bit. If fitted incorrectly the drop can be uncomfortable for the horse – if too low it will rest on the soft skin of the nasal passages and can interfere with the horse's breathing.

Flash noseband

This has an upper strap which fits in the same way as a cavesson but is tightened rather than left loose. It has a second strap attached to the cavesson part, which fastens around and under the bit. This noseband is more effective at preventing the horse from crossing his jaw and opening his mouth than the drop and is in very common use. The way in which it fits makes it more comfortable for the horse (it does not end up over the soft skin of the nasal passages).

Grakle noseband

This noseband, named after the winner of the 1931 Grand National and often misspelt Grackle, consists of two straps which cross over on the front of the horse's nasal bone and fasten tightly in the same position as the Flash. This can be adjusted fairly precisely as required and is useful on strong horses when going cross-country jumping. As it acts over a wider area of the head it prevents the horse from crossing his jaws.

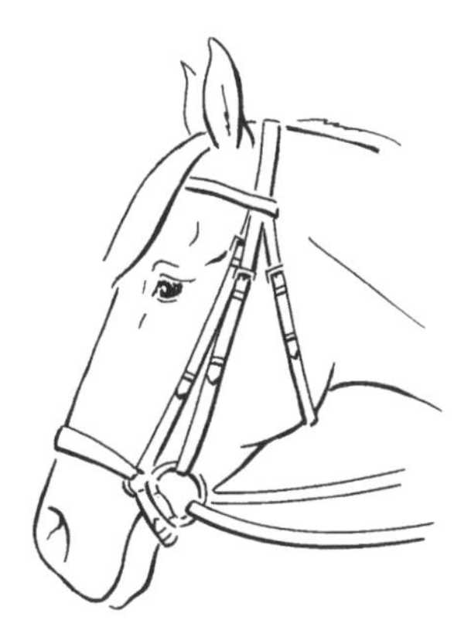

3.2 The drop noseband

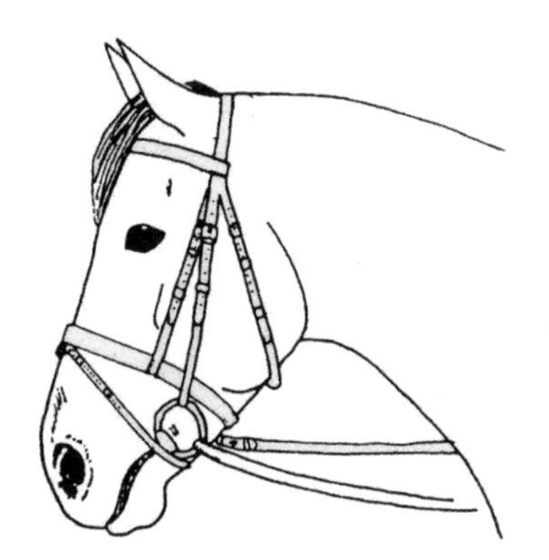

3.3 The Flash noseband

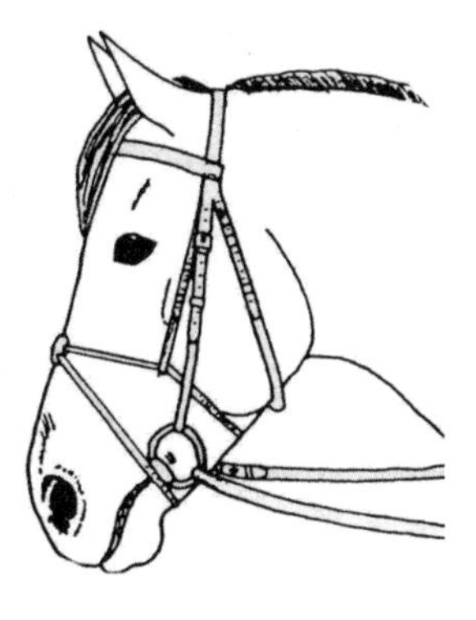

3.4 The Grakle noseband

Types of rein

Different types and length of **rein** are available. Reins must be long enough to allow the horse to stretch when jumping, but not too long, as the rider's foot may become entangled in them. The width of the rein depends on the size of the rider's hands.

There are several materials used for reins, including:

Rubber-covered reins. These offer good grip in all conditions. Once the rubber has worn out, they can be re-covered. Rubber-covered reins are the most frequently used and are essential for any form of cross-country riding.

Plain leather. These are smart for showing but do not offer much grip, especially when wet, or when too much saddle soap has been applied.

Plaited leather. These offer a slightly better grip than plain leather, but can be uncomfortable to hold if allowed to get dried out.

ITQ 3.1

Label the diagram:

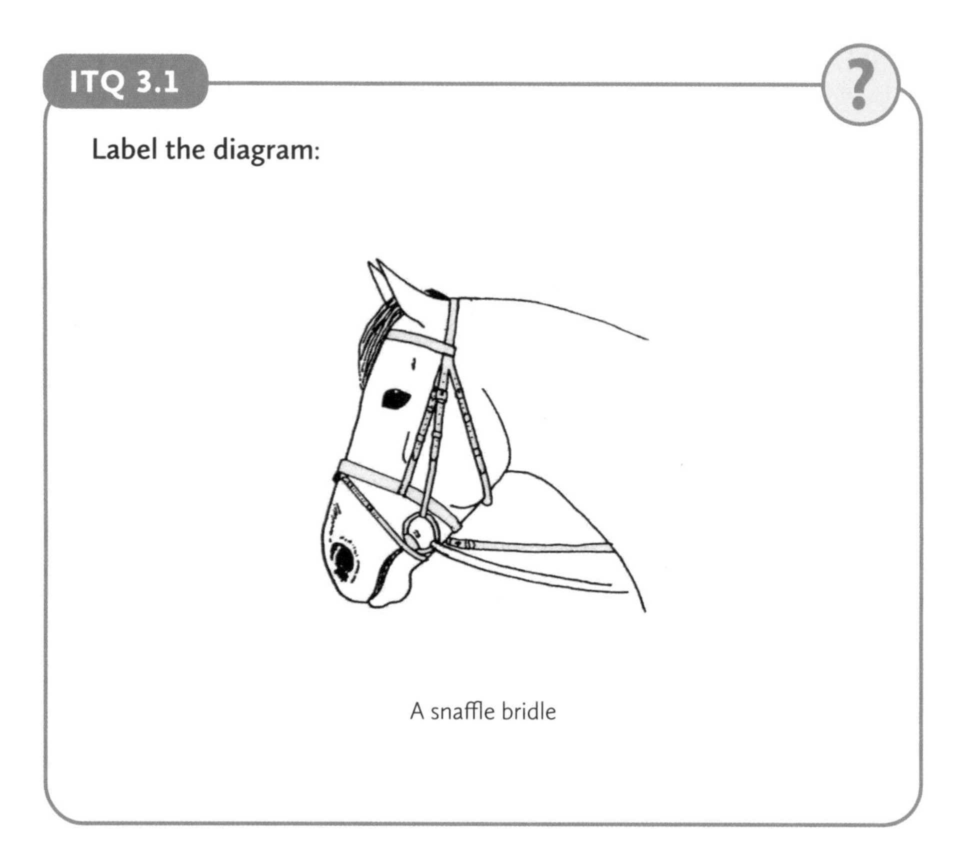

A snaffle bridle

ITQ 3.2

How can you tell if a cavesson noseband is correctly fitted?

ITQ 3.3

What is the main purpose of the Flash noseband?

ITQ 3.4

How can you tell if a snaffle bit is fitted correctly?

Continental reins. These are made of webbing and have leather grips at intervals, so are good for ensuring the rider has even hold of both reins, but are not easy to make minor adjustments with.

Bits

Various patterns of **snaffle** are the most commonly used bits; the basic patterns are generally suitable for novice riders.

Single-jointed snaffles

Single-jointed snaffles have one joint in the mouthpiece which gives the bit a **'nutcracker'** action when the joint closes, i.e. as pressure is applied to the reins. It also acts on the bars and lips. There is less tongue pressure than with a straight-bar snaffle.

Single-jointed snaffles are further defined by the pattern of the bit rings.

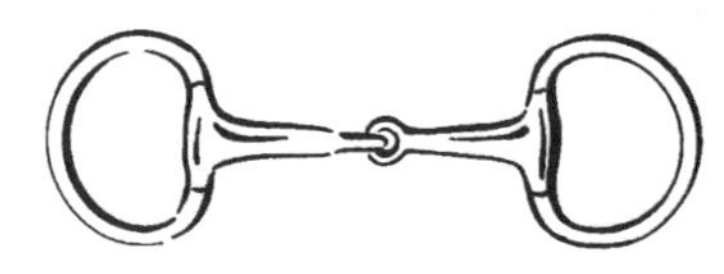

Eggbutt snaffle

Loose-ring snaffle

3.5 Single-jointed snaffles

Loose-ring snaffle. This bit encourages the horse to relax his jaw and salivate (known as **mouthing**) and is widely used. Care must be taken that the skin on the sides of the mouth does not get pinched in the loose rings. If the holes in the mouthpiece, through which the bit rings pass, become worn and enlarged, the bit must not be used as this increases the risk of nipping the skin. The **German loose-ring hollow mouth** is a very light version of this bit.

Eggbutt snaffle. This very popular bit is acceptable to most horses. The bit rings do not allow the mouthpiece to move as much as the loose-ring so it is useful for horses with a tendency to have 'fidgety' mouths. The bit is also less prone to nipping as it does not pull through the mouth so easily. The thickness and weight of the mouthpiece varies. A thicker mouthpiece is essentially milder in action than a thinner one because the pressure of the reins is spread over a wider surface area (but it may not necessarily be more *comfortable* if the horse has a very small mouth.) As with the loose-ring version, **German eggbutts** are hollow, so therefore very light.

Another variation of the eggbutt is the **'D' ring** which, as the name suggests, has 'D' shaped bit rings.

Double-jointed snaffles

The 'nutcracker' action of a double-jointed snaffle is far less marked than that of a single-jointed one, the main action being on the bars of the mouth and the lips. As with single-jointed snaffles, double-jointed snaffles can be either loose-ring or eggbutt.

Double-jointed snaffles can have different patterns of mouthpiece.

French link

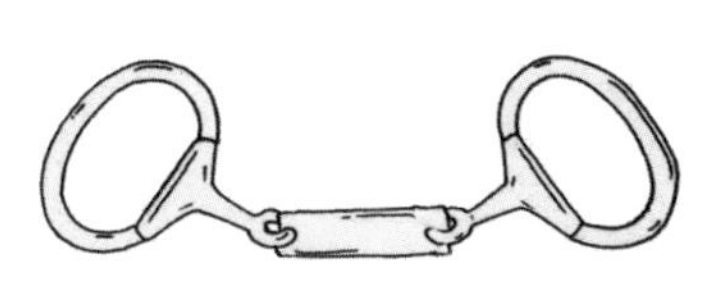

Dr Bristol

3.6 Double-jointed snaffles

French link. This has a rounded plate in the centre of the mouthpiece, which lies on the tongue. It encourages the horse to relax his jaw. It must be the correct fit for the horse and not pull through his mouth when pressure is applied to one rein. If it pulls through, the joint with the centre link will nip the corner of the lips.

Dr Bristol. This bit has an oblong centre plate which rests on the tongue. The plate has squared edges and should always rest flat on the tongue. It is a severe bit, used on strong horses.

3.7 Straight-bar mullen-mouth snaffle

Straight-bar snaffles

Straight-bar bits have an unbroken mouthpiece and act mainly on the tongue and, to a lesser extent, the bars of the mouth. Many are of a **half-moon** or **mullen-mouth** shape. As these are shaped to fit the contours of the mouth they are mild in their action. Some common patterns of mouthpiece for these snaffles are as follows.

Vulcanite loose-ring. Vulcanite is a very hard, unyielding material which does not encourage the horse to mouth the bit. Vulcanite mouthpieces can be rather bulky as they are not shaped.

Nylon. These are often shaped and provide a hard-wearing, flexible, mouthpiece which encourages the horse to mouth and accept the bit. They are ideal for young horses and those who are particularly sensitive. (Some jointed bits are now made of this material.)

Rubber. Rubber mouthpieces have the same qualities as the nylon ones although they are not normally shaped. Horses tend to chew excessively on rubber bits, which may lead to wearing.

Metal loose-ring. Straight-bar metal mouthpieces are not usually used for riding purposes but may be used on stallion bridles (for leading the stallion).

THE SADDLE

Types

Saddles can be divided into three main types:

Dressage. The dressage saddle is a specialist saddle used only for flatwork. It is designed to allow a longer length of leg as the panels are very straight cut. A short **Lonsdale girth** is used as the dressage saddle has extra long girth straps.

Jumping. Again a specialist saddle with very forward-cut panels and large knee rolls for the shorter leg position used when jumping. The knee and thigh rolls are designed to help the rider maintain the leg position.

General-purpose. This is the most commonly used type of saddle. It is a compromise between the above two and is suitable for use in the basic levels of flatwork and jumping, and for general hacking. You will use a general-purpose saddle in your Stage 1 exam.

There are also saddles made for other specific uses, such a racing, polo, endurance and Western-style riding, but these are beyond the scope of this text.

Saddle structure

The **tree** is the framework upon which the saddle is made. The majority of trees used to be made from beech wood, but more recently laminated wood has been used, which can be moulded into shape to give a strong, lightweight tree. Also, some trees are now made of high-tech artificial materials. Trees are generally made in three widths – narrow, medium and wide, (although some are now adjustable) and in different lengths.

The tree can be **rigid** or **'sprung'**. The **spring tree** consists of two flat panels of steel running from underneath the pommel (front) to the cantle (back). The panels are thin enough to give a 'springy' feel to the tree. The rigid tree has a more solid framework than the spring tree.

The **stirrup bars**, made of hand-forged steel, are riveted to the **points** of the tree. The stirrup bars are often recessed and have a safety catch *which should always be in the open position when in use*. This ensures that, in the event of the rider falling and a foot becoming trapped in the stirrup, the whole stirrup leather will come off the bar. The catch is only closed when the saddle is being transported or used on a horse who is being lunged without a rider. However, because of the recessing, it is not normally necessary to put the safety catches up in these circumstances, as the leathers are often extremely difficult to put on and take off, especially on new saddles.

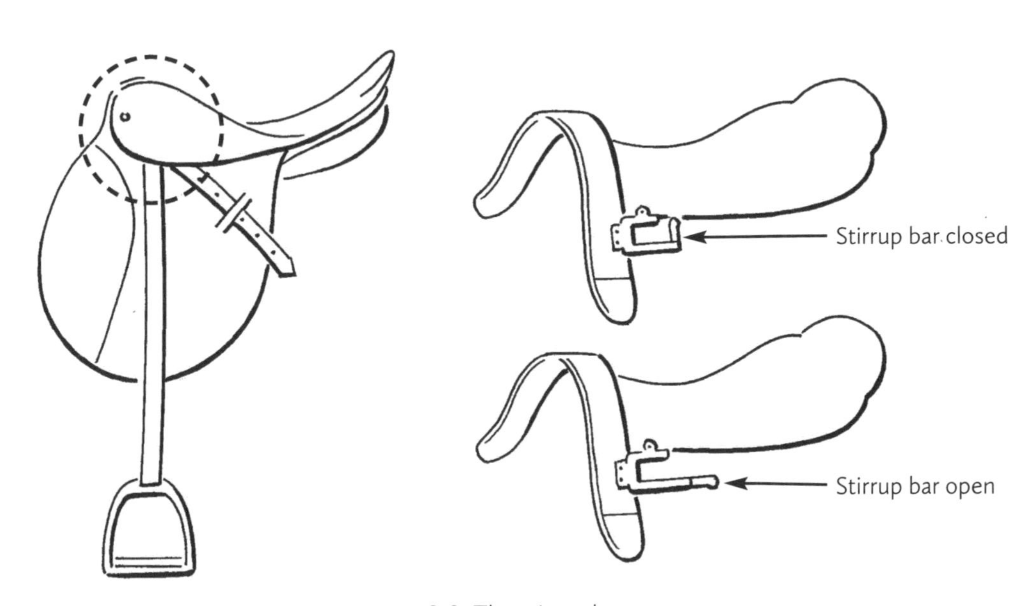

3.8 The stirrup bar

Saddles usually have full **panels**, extending down the same length as the flaps, although some pony saddles only have half panels. Full panels give a greater bearing surface which is more comfortable for the horse.

Traditionally, the panels are stuffed with wool fibre, known as **'flocking'** (although some modern designs have air-filled panels). Saddles need regular attention (at least once a year), as the flocking packs down, often unevenly – the saddler will put this right. If not corrected, uneven flocking can cause pressure points, which will give the horse a sore back.

There are normally three **girth straps** on each side of the saddle. It is correct to use the front two straps, keeping the third straps as spares in case of breakages. The front straps are attached to a separate piece of webbing whilst the second and third straps are connected to the same piece of webbing. To use the second and third girth straps together would increase the risk, should the webbing break, therefore the second and third straps should never be used together. Girth straps should be checked frequently for wear and tear, especially the stitching.

Leather **buckle guards** should always be used to prevent the metal girth buckles rubbing and wearing through the saddle flap. New saddles tend to have built-in buckle guards.

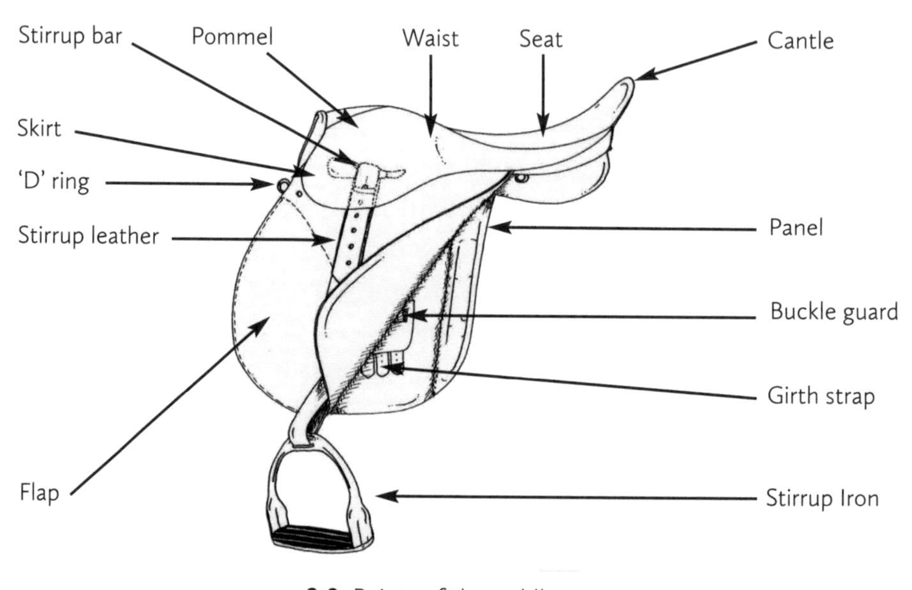

3.9 Points of the saddle

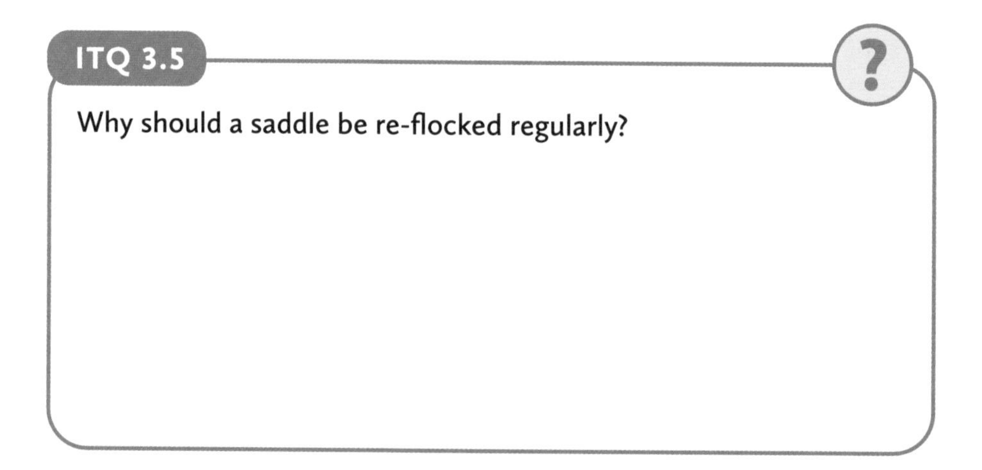

Saddle fitting

It is important that the saddle is a good fit for both the horse and rider. Serious damage can be caused to the horse's spine through using an ill-fitting saddle. In the Stage 1 exam the saddle should fit the horse, but you should be able to recognise signs of a good fit.

If a saddle fits the horse badly, he may show his discomfort by behaviour such as bucking, refusing to jump and generally poor performance. Very often though, a horse will continue to work properly, even though he is in pain. Therefore it is up to us to make sure that only well-fitting saddles are used.

- When trying a new saddle to see if it fits, make sure that the horse is clean and place a clean stable rubber on his back to stop the saddle from being marked. Do not use a numnah as this disguises the true fit of the saddle. Use buckle guards to prevent the flaps from being marked by the girth. Fasten the girth.
- The saddle should sit level on the horse's back – it should not tilt forwards or backwards as this would make it difficult for a rider to maintain position and would exert uneven pressure on the horse's back.
- The saddle lies on the lumbar muscles, which cover the top of the ribs. It should rest evenly on these muscles and must not touch the loins. The full surface of the panels must be in contact with the horse's back to distribute the weight over the largest possible area.
- The saddle must not pinch the shoulders. The knee rolls, panels and saddle flaps should not extend out over the shoulders as this would restrict the horse's freedom of movement.
- With the rider mounted there should be sufficient and consistent clearance beneath the pommel when in an upright, flatwork position and a forward, jumping position.
- You should be able to see daylight through the gullet. At no point or time should the saddle touch the horse's spine. When viewed from behind the saddle should not appear crooked or twisted.
- If buying a saddle, you must walk, trot and canter in it (with stirrups attached) to see how it feels. It is not sufficient to just sit in it on the yard.

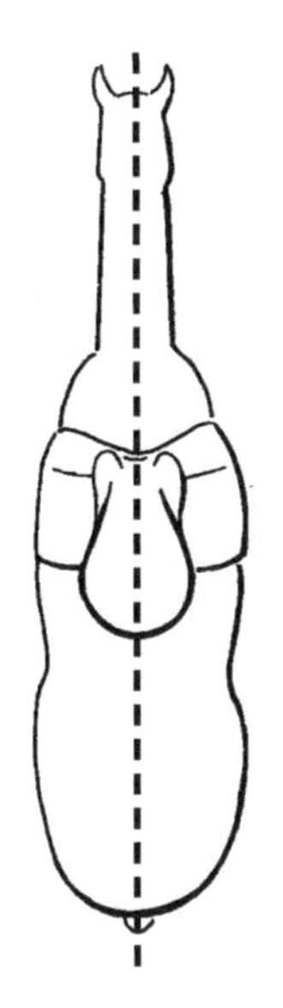

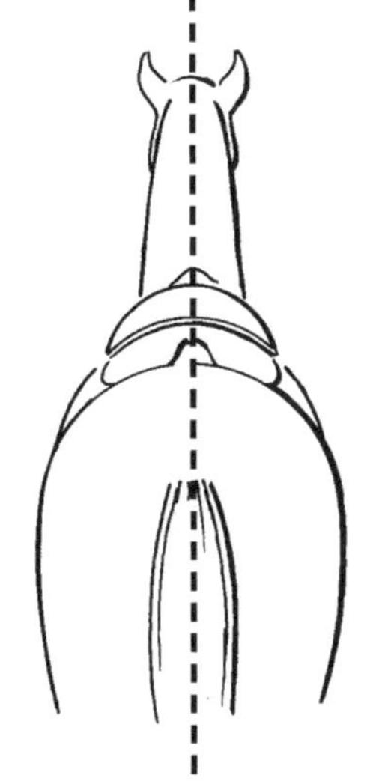

When viewed from behind the saddle should be straight and not twisted

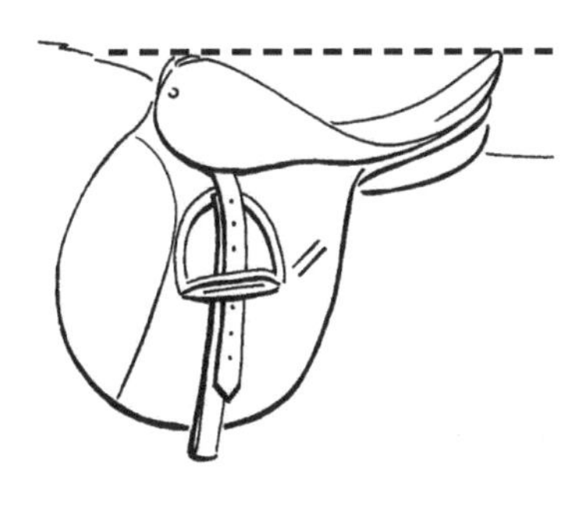

The pommel and cantle should be level

3.10 Saddle fitting points

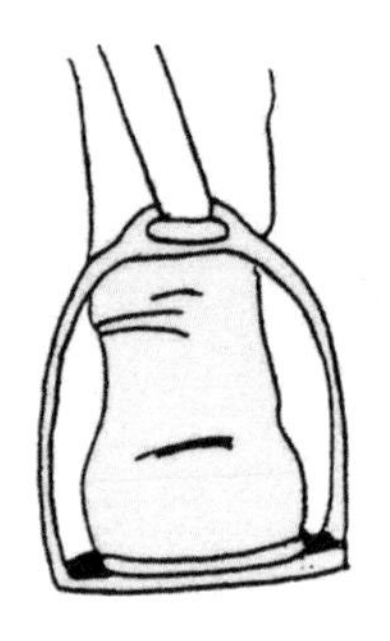

Correct fit of stirrup

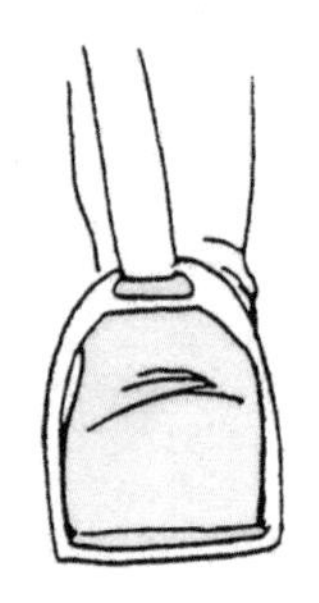

Stirrup too small

3.11 Stirrup fittings

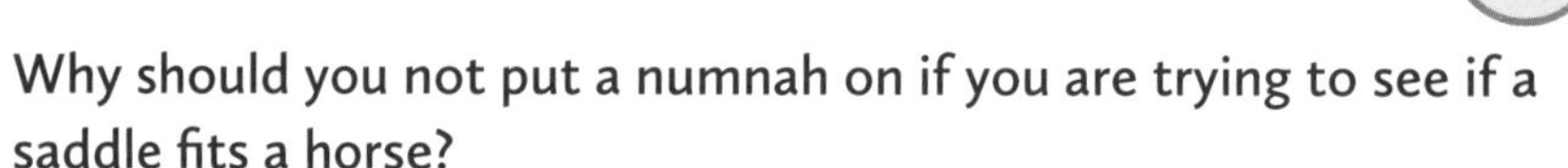

ITQ 3.6

Why should you not put a numnah on if you are trying to see if a saddle fits a horse?

Stirrup irons

The stirrup irons must be the correct size for the rider – neither too large nor too small. If too large, the rider's foot can slip right through and become trapped; if too small, the foot can become wedged in the iron.

HANDLING SADDLES AND BRIDLES

Tack is expensive and can be ruined through careless handling.

- Never leave tack within the reach of a horse. Most horses will have a chew or, if tack is on the ground, they may tread on it.
- Saddles must not be left balanced on the bottom stable door; some horses take great delight in knocking saddles down onto the concrete yard.
- When handling the saddle always ensure that the stirrups are run up. If you have to put the saddle down on the ground, rest it on the pommel using a numnah to protect the saddle from scratches. Place the girth over the cantle and rest this on the wall. Never allow the saddle to touch the concrete or brick walls/floors as the leather will scratch and eventually wear through completely.
- If the saddle is dropped or rolled on, the tree could break. To test for a broken tree put the front arch on your knee and try to move the cantle towards the pommel. Then apply pressure inwards to both panels at the front. Any extra movement or grating sounds would indicate a broken tree. Under no circumstances should a saddle with a broken tree be used as it would cause damage to the horse's back.
- Never leave bridles lying around as people or horses could become entangled. Always rinse the bit immediately after use and hang the bridle up on a cleaning hook ready to be cleaned before being hung up neatly. Do not allow reins, girths, etc. to trail along the ground when carrying tack.

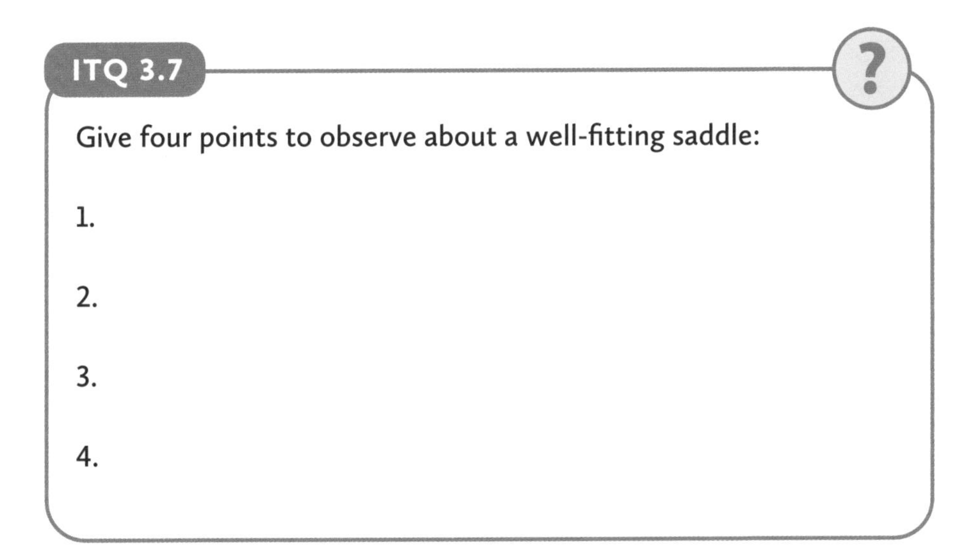
ITQ 3.7

Give four points to observe about a well-fitting saddle:

1.

2.

3.

4.

OTHER ITEMS OF TACK

Numnahs

A numnah is a soft pad used beneath the saddle, made from a variety of materials including sheepskin, cotton-covered foam and neoprene. A numnah is used to help prevent the saddle chafing the horse's back. It helps by absorbing sweat and absorbing and evening out the pressure caused by the rider's weight. Numnahs are *not* used to make an ill-fitting saddle fit.

Thin-skinned horses are sometimes very sensitive to the cold leather of the saddle. Numnahs are a great help to 'cold-backed' horses. Numnahs must be kept well brushed to prevent hair from balling up and causing lumps, and should be laundered regularly.

To fit correctly, they should be pulled well up into the arch of the pommel. The numnah must never rest flat on the horse's back as it will rub as he moves and cause pressure sores. The numnah should protrude evenly all the way around the saddle. All fastening straps should be correctly fitted around the girth and adjusted to stop the numnah slipping out backwards from beneath the saddle.

Girths

There is a variety of different types of girth and their sizes range to fit every size of horse or pony. Using the right size is important. Girths that only 'fit' if on the first or last holes of the girth straps are unsatisfactory and potentially dangerous. It is important that the stitching and buckles are sound and that they are kept well cleaned to prevent girth galls, i.e. sores on the horse's sides.

Girths are made from neoprene, leather, cotton, webbing, nylon and padded synthetic materials, with or without elastic inserts.

Leather girths

These are expensive but are very hard-wearing. They should be cleaned regularly, kept soft and supple to prevent chafing and regular attention paid to the condition of the stitching. The most popular types of leather girth are the Atherstone, the Balding and the threefold.

- **The Atherstone** is slightly padded and shaped behind the elbows to prevent galling.
- **The Balding** consists of three strips of leather fastened together in such a way as to prevent galling.
- **The threefold** consists of one piece of leather folded over in three. The rounded edge of the fold goes towards the horse's elbows to prevent rubbing. Care must be taken to keep the girth clean in the folds, and supple.

Another type of girth is the **Lonsdale** – this is short girth, used on a dressage saddle which has long girth straps. The buckles, being much lower, do not interfere with the rider's legs but are difficult if not impossible to adjust by a mounted rider. Lonsdale girths may be made from leather or a synthetic material.

Leather girths do become worn and the stitching will rot. Only relatively new leather girths should be used when jumping as there is a risk that old leather girths will break under duress.

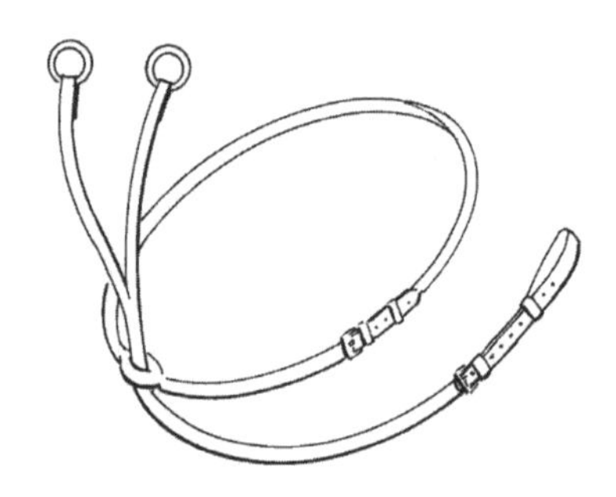

3.12 The running martingale

Running martingale

Martingales are artificial aids used to prevent the horse's head being raised so high that the rider has no control. They come into action once the head has been raised, at which point the horse feels pressure on his mouth: they are not used to force the horse's head down.

On a running martingale the reins run through two rings attached to straps which fasten around the girth. A neckstrap holds this in place. When fitting a running martingale the neckstrap buckle goes on the left and should be adjusted to admit a clenched fist.

To check the fitting of a running martingale, when the horse stands with his head in the normal position the martingale straps must be slack. It should come into action only when the horse raises his head beyond the point of control. Before passing the reins through the martingale rings, take both straps and hold the rings together towards the horse's withers – the rings should not touch the withers as this would make the fit too loose. Then pass the reins through the rings and re-buckle; standing beside the horse, hold the reins up in the position they would be in if the rider were holding them on the horse. Check the fit – the straps should be slightly loose. Make adjustments on the buckle which lies between the forelegs. Always ensure that this buckle faces away from the skin to prevent chafing.

The neckstrap should be adjusted to admit the width of your hand. A rubber stopper must be used on the neckstrap to prevent it slipping along the main strap, which will affect its action. Rein stops are used on each rein to prevent the martingale rings running up too near the bit rings, which could panic a horse.

The hunting breastplate

SAFETY TIP

- The girth must be fastened before the straps are attached to the 'D' rings of the saddle. If you attach the 'D' rings first, i.e. before you fasten the girth, and the saddle slips, it will end up entangled around the horse.

The hunting breastplate is used to prevent the saddle from slipping backwards, particularly in competition circumstances, e.g. cross-country jumping. It consists of a leather strap that passes between the horse's forelegs, through which passes the girth. Attached to a ring at the breast is a V-shaped neckstrap that is attached by straps and buckles to the 'D' rings either side of the pommel. A martingale attachment can be attached to a hunting breastplate.

It should be fitted just before the saddle, as the girth must pass through the breastplate strap.

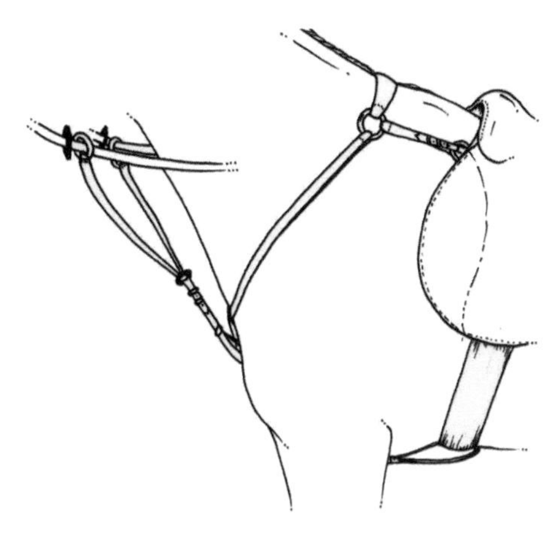

3.13 The hunting breastplate

ITQ 3.8

Describe how you would put a saddle on the ground if you had to.

ITQ 3.9

What are the purposes of using a numnah?

ITQ 3.10

What is the purpose of using a running martingale?

ITQ 3.11

How can you judge the correct fit of a running martingale?

TACKING UP

Clothing/equipment notes:

Take off your gloves and put them in your pocket when tacking up – you cannot adjust buckles or straps wearing gloves.

Preliminaries

EXAM TIP

If the reins do not have rein stops, you must comment on this in your exam.

Before tacking up, the horse should be tied up safely and groomed. If brushing boots are worn these can be put on first.

If a martingale is being used, unclip the lead-rope from the headcollar and slip the martingale over the horse's head just before you put the saddle on. (This assumes that the martingale is not attached to the bridle.) Re-clip the lead-rope onto the headcollar.

If the martingale is attached to the bridle, put the bridle on before the saddle. Make sure the reins have martingale stops on.

Saddling

1. The horse may be tied up in the yard or the stable. When learning to tack up it is safest to have the horse in an enclosed area in case he gets away from you. It is always safer to have the horse tied up whilst putting on the saddle as it prevents him from wandering around with the saddle balanced on his back.

2. The girth is likely to be buckled on the offside (check this) and should lie over the top of the saddle. The stirrups must be run up the leathers to prevent them banging on the horse's legs and sides.

3. Standing on the nearside (left) place the saddle on the horse's back, forward over the withers. The numnah or saddle cloth must be pulled up into the gullet under the arch of the pommel to prevent it pressing down on the spine. Slide the saddle down into position and check that the saddle flaps are not folded back.

4. Check the saddle on the offside (right) for any creases and bring the girth down so it lies just behind the horse's elbow.

5. Make sure that all numnah fastenings are secure and pass the girth through the appropriate straps. Most numnahs have straps for the girth to pass through. Some have Velcro so are easy to adjust.

6. Return to the nearside and fasten the girth. If you are using a martingale or hunting breastplate, first thread the girth through the martingale/breastplate strap. The girth should sit immediately behind the horse's elbow and should not be fastened too tightly. Make sure the skin at the elbows is not pinched by the girth. You may need to straighten it with your hand.

Fitting the bridle

1. As the bridle is likely to have been 'put up' after its last use you will need to undo the throatlash and noseband. Standing on the nearside, undo the quick-release knot of the lead-rope, leaving the rope through the weak link, and slip the headcollar around the horse's neck and do it up. This will give you some control of him.

SAFETY TIP

- The rope must be untied to prevent the headcollar from 'strangling' the horse in the event that he should pull back.

2. Pass the reins over the horse's head and neck. Facing forwards, hold the bridle by the cheekpieces in one hand, whilst controlling his head with a gentle pressure

3.14 Putting on the bridle

on the front of his nasal bone. Some horses will be uncooperative and throw their heads up, so it is important to have your hand in position to control this.

3. Using your other hand, guide the bit into the horse's mouth. If he does not willingly open his mouth, you may need to apply pressure to the corners of the lips to encourage him to do so.
4. Once he has opened his mouth, slip the bit in and raise the bridle to prevent the bit from dropping out of the horse's mouth.
5. Gently pass the headpiece over the horse's ears, making sure that the ears are not pinched by the browband. Check that the forelock and mane are disentangled from the bridle and smoothed down.
6. Fasten the throatlash, allowing the width of a clenched fist between it and the throat area.
7. Ensure the bridle is on straight and that the bit is level by standing in front of the horse and making adjustments where necessary.
8. Fasten the noseband, according to the type used.
9. If a running martingale is used and it is not yet attached to the bridle, undo the buckle on the end of each rein and, without dropping either rein, pass each through its respective martingale ring making sure neither the rein nor ring is twisted. Refasten the buckles.
10. Ensure all straps are tucked neatly into their keepers.
11. If the horse is to be left tacked up in the stable, at Step 6 above, you would secure the reins by twisting them around each other, taking up all slack and passing the throatlash through them and fastening. This prevents the reins from hanging down low. The headcollar must then be put on over the top of the bridle and the horse tied up.

3.15 Reins made safe – two methods

UNTACKING

Removing the bridle

1. On dismounting, run up your stirrups immediately to prevent them from banging around. If leading the horse back to the yard, you should loosen the girth by one or two holes and take the reins forward over his head. Once back in the stable or yard, put the reins back over his neck.

2. Make the headcollar ready to use, i.e. unbuckle it and remove it from the ring where you left it hanging safely. Place the lead-rope around the horse's neck so you have control.

3. Undo the noseband first, then the throatlash.

4. Holding the headpiece, gently pull the bridle off forwards over the horse's head, allowing the bit to drop out of his mouth.

5. Put the headcollar on the horse.

6. Bring the reins forwards again over his neck so the bridle is now fully removed.

7. Tie the horse up.

8. In normal circumstances you must rinse the bit to prevent the saliva from drying before you hang the bridle up. You may not have the opportunity to do this in the exam.

If the horse is wearing a martingale you will need to put the headcollar on over the bridle, tie the horse up and remove the saddle first as the girth passes through the martingale loop. Then remove the martingale with the bridle.

Unsaddling

1. On the nearside, undo the girth.

2. If a martingale or breastplate is used make sure that you pull the girth out from the martingale/breastplate strap. Grasp the pommel in your left hand and the cantle in your right. Lift the saddle and numnah off the horse's back.

3. Put your left arm under the gullet so you can carry the saddle on the crook of your arm. Use your right hand to catch the girth and pass the girth over the top of the saddle.

SAFETY TIP

- If a hunting breastplate is used you must undo the breastplate strap buckles and remove the straps from the 'D' rings before you undo the girth.

4. Place the saddle safely where it cannot be trodden on or knocked over, preferably on a saddle horse or rack.

5. Brush the horse's back off to remove saddle marks. If the weather is hot you may need to sponge off sweat marks. As part of your daily routine this is the ideal time to strap the horse as the pores of the skin will be open after exercise. He should always at least be brushed off and have his feet picked out.

Now is also the time to remove brushing boots or any other protective legwear.

TACK CLEANING

In your exam you are likely to be asked about tack cleaning rather than having to demonstrate it.

Reasons

It is essential to clean tack regularly for the following reasons:

- Cleaning keeps the leather supple and less likely to crack.
- Supple leather is more comfortable for the horse and rider.
- Cleaning gives you the chance to inspect the tack for safety.
- Clean tack is a pleasure to look at and work with. Dirty, stiff leather and numnahs are unpleasant to work with, show a lack of attention and care, as well as being potentially dangerous.

Equipment

- Tack-cleaning hook for hanging all the straps, leathers and girth.
- Saddle horse.
- Two buckets of clean, warm water.
- At least two sponges, one of which must be kept dry.
- Clean cloths (tea towels).
- Metal polish and cloths.
- Saddle soap (glycerine bars are ideal).

We'll look first at how tack should be kept clean on a daily basis, and then how to give a thorough cleaning.

The daily clean

Tack should be wiped over as soon as you finish with it. This is a good habit to get into as it is a lot easier to clean tack that is still a bit warm from use before any sweat and grease dries and hardens. As mentioned earlier, the bit should be rinsed as soon as you remove the bridle.

1. Put the saddle on a saddle horse (it should always be cleaned on a saddle horse or rack for ease and safety) and hang the bridle on the cleaning hook. Pull the straps from their keepers but do not dismantle the tack.

1. Wring out a sponge in warm water. The sponge must not be too wet as it will make the leather soggy and dull. Clean off all grease from the leather – you will need to rub fairly hard on the underside of the leather (the side that is in contact with the horse's coat). The upper side is fairly easy to get clean. Keep cleaning the sponge in the warm water, remembering to wring it out really well each time.

3. You may see small black spots on the tack – these are called **'jockeys'** and need to be removed. If they won't come off with the sponge, you can carefully scrape them off with a blunt knife, taking care not to scratch the leather.

4. If there are plugs of grease in the holes of the straps, press them out using a matchstick.

5. Having removed all grease, rub the leather over with a dry tea towel. This removes the last traces of grease and any residual water.

6. Now take your dry sponge. Dip the end of the saddle soap bar in the warm water and rub onto the sponge. Don't put the sponge into the water or you will end up with lather, which makes the leather damp and dull. Apply the saddle soap over both sides of the leather. Keep rubbing the bar onto the sponge, remembering to keep the sponge dry. The leather should have a shiny appearance if you are doing it correctly. It takes practice to soap leather and make it shine. If the sponge is too wet the leather will look dull and soggy.

7. Wipe the stirrup irons and bit with a clean cloth.

8. Replace all the bridle straps in their keepers and, holding the reins at the buckle (a pair of reins is held together by a buckle in the middle), pass the throatlash through the reins and do it up. Fasten the noseband around the outside of the bridle and hang up.

9. Brush loose hair from the numnah and, if it is damp, hang it outside or on a radiator to dry.

ITQ 3.12

Give three reasons for cleaning tack regularly:

1.

2.

3.

The thorough clean

Occasionally tack will need a thorough clean.

1. To clean tack thoroughly it should be dismantled. Undo all buckles, making a note (either in your head or written) of which holes the buckles were on. It is not really necessary to remove the rubber stirrup grips from the irons unless you absolutely have to – they can be difficult to get back in again. Hang all the straps on the cleaning hook.

2. If the stirrup irons are very muddy, place them in a separate bucket of warm water with the bit and then clean and dry them. Stainless steel irons will clean up without the use of polish but if you are preparing for a show, e.g. a turnout class or similar, you may want to use metal polish. *Don't* use metal polish on the mouthpiece of the bit.

3. Clean the leather in the method previously described.

4. Now you have to reassemble the tack. When you are learning, this can be the interesting bit. You put it all back together again and find that you have one piece left over! To reassemble a bridle:

   - Start with the headpiece and thread the browband onto it. Imagine how it looks on the horse as you do this. Think of where the ears come through and that the throatlash does up on the left. One common mistake is to have the headpiece the wrong way around. Now hang the headpiece on the cleaning hook.
   - Next, attach the cheekpieces to the headpiece. If they have billets they go to the inside; buckles fasten on the outside. Remember to put them back on the correct holes. As they hold the bit in place, it is very important that they are in the correct position.
   - Attach the bit to the cheekpieces, making sure it is the correct way up – i.e. it should hang in a smooth, rounded shape.

- Now attach the noseband. Make sure the noseband is the right way round (again, imagine how it would look on the horse). Thread the headpiece of the noseband under the right side of the browband, underneath the main bridle headpiece, down through the left side of the browband and buckle up on the correct hole on the left-hand side.

- Attach the reins to the bit. Remember – billets to the inside, buckles to the outside. Pass the throatlash through the reins as previously described, fasten and do up the noseband before hanging up the bridle.

Oiling

Occasionally, tack may need to be softened with oil. This can be done four to five times a year to keep it supple. Clean to remove grease and then sparingly apply a *light* coating of a suitable oil. There is a wide range of leather dressings available from tack shops. Once the oil has been fully absorbed, apply the saddle soap. Don't apply oil too often as this can lead to the tack becoming soggy and limp. Don't overdo the oil on the seat of the saddle as it will stain jodhpurs.

SAFETY CHECKS

Check the following regularly:

- The condition of all stitching. Any repairs must be carried out promptly to avoid an accident. Never use tack that is in need of repair or that is *very* old. A broken rein whilst cantering in an open space can be disastrous.

- On the saddle, keep a close eye on the stitching that holds the girth straps onto the webbing under the saddle flaps. Check the webbing for signs of wear.

- Leather girths can wear and, under periods of strain, e.g. jumping, may break. Very old leather girths should be replaced. They *definitely should not* be used for showjumping or cross-country riding.

- Check the stitching on the stirrup leathers. Also, if one rider uses the leathers all the time (i.e. at the same length), they may start to show signs of wear where the leather is constantly bent. It is also sensible to switch leathers between nearside and offside, since the nearside tends to take the extra strain of mounting.

- All buckle holes, especially on girth straps, stretch eventually and can wear to such an extent that one hole runs into another.

- If the stuffing in the saddle starts to feel lumpy and uneven, consult the saddler, who will pull out the old stuffing and re-flock the saddle. Once this has been done, the fit can be re-checked.

ITQ 3.13

When cleaning tack you should check for signs of wear. Give three areas which need to be checked regularly:

1.

2.

3.

Injuries caused through dirty or ill-fitting saddlery

Saddle sores

Symptoms

- Swelling on the back in the saddle area.
- The skin may be broken and sore.

Causes

- A horse who has been rested for a long period and just brought back to work will be in 'soft' condition, i.e. overweight, with slack muscles. At this stage the skin is prone to rubbing.
- Friction and uneven pressure caused by ill-fitting saddles and/or bad riding.
- Friction caused through lack of cleanliness, e.g. ungroomed coat or dirty numnah.
- Thin-skinned horses can be prone to rubbing.

Treatment

- Remove the cause.
- Stop all work under saddle, although the horse may be exercised in hand or on the lunge. Remember to adjust feeding as appropriate.
- Treat broken skin with warm salt water. This has mild antiseptic qualities and will also help to harden the skin.
- If the skin is broken, check that the horse is up to date with his tetanus vaccinations.
- Do not use a saddle until completely healed.

Prevention

- Always ensure that the horse is clean before tacking up.
- Use good quality numnahs and make sure they are regularly brushed and washed.
- Saddles require regular checking by a saddler as the flocking used to stuff them will shift and cause uneven pressure. Most saddles need to be re-flocked annually.
- When bringing a horse in soft condition back into work, apply surgical spirit to the saddle and girth areas to harden the skin.
- After exercise, remove sweat marks by sponging and/or grooming

Girth galls

Girth galls are similar in every respect to saddle sores. These are found usually just behind the elbow, especially when nylon string girths are used. Thin-skinned horses are prone to galling.

Causes and treatment
As for saddle sores. If a horse is prone to galling a sheepskin girth cover may be used.

Bit sores

Symptoms

- These vary from slight bruising to bleeding sores on the corners of the mouth.
- A horse with a sore mouth will not be receptive to rein aids.

Causes

- An ill-fitting or badly adjusted bit.
- Too severe a bit in use.
- If a horse is very strong to ride and constantly pulling.
- Bad riding by a novice or ignorant rider.

Treatment

- Remove the cause.
- Rest the horse until the mouth is healed. Exercise in hand or on the lunge, without using a bridle. A lunge cavesson should give sufficient control.
- Exercise in a bitless bridle, e.g. a hackamore. (An inexperienced rider must not use a hackamore as its action is severe if used incorrectly).

- Apply warm salt water to the corners of the mouth.
- Once the mouth has healed, apply petroleum jelly to the corners whenever a bit is used. Use a gentle form of bit, e.g. rubber snaffle.
- Some products on sale for human mouth conditions have valuable healing properties and can be useful for horses.
- If the problem is caused by bad riding, educate the rider and encourage the use of a neckstrap. A good riding instructor should be sought.

ITQ 3.14

Give two causes of saddle sores:

1.

2.

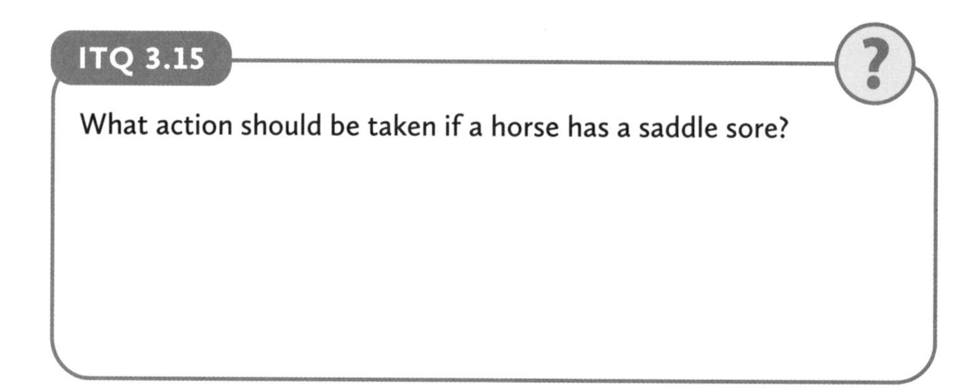

ITQ 3.15

What action should be taken if a horse has a saddle sore?

4 Horse Husbandry – Bedding and Mucking Out

REQUIRED SKILLS/KNOWLEDGE	Learnt, revised, practised?	Confirmed
Skip out and set fair a bed.		
• Use stable tools safely and efficiently.	☐	☐
• Skip out efficiently and set fair a bed.	☐	☐
Demonstrate knowledge about bedding.		
• Discuss different bedding materials and understand their various uses.	☐	☐
• Know how to maintain different types of bedding.	☐	☐
• Know how to muck out and bed down correctly.	☐	☐
• Describe how to store and dispose of stable waste.	☐	☐

BEDDING

Bedding is essential for the following reasons:

- It encourages the horse to lie down and rest.
- It prevents the horse from slipping on the floor.
- It prevents the horse's legs from getting jarred whilst standing in the stable.
- It encourages the horse to stale (urinate).
- It provides warmth and excludes draughts.

It also looks aesthetically pleasing.

Qualities and types of bedding materials

Bedding materials should have the following qualities:

- Non-harmful if eaten.
- Warm and comfortable.
- Non-irritating.
- Dust- and mould-free.
- Absorbent or draining.
- Non-slip.
- Easily obtainable.
- Disposable.

Choice of bedding material

Your choice of bedding will depend on the following:

- **Availability**. What type is the most easily obtained in your area/part of the world? In areas where straw is in short supply, alternatives such as shavings or paper will have to be used.
- **Cost**. What is your budget? Straw is usually the least expensive type of bedding. Dust-extracted straw is slightly more expensive whilst shavings and paper are the most expensive forms of bedding.
- **Storage.** Do you have a barn? Polythene wrapped bedding material can be stored outside although it is desirable to keep it under cover. Dust-extracted straw is packed in polythene but ordinary straw will need to be stored in a dry barn. Outside stacks covered in a tarpaulin are not very successful as the outer bales get ruined, the stack eventually collapses and you are left with a large pile of wet, rotten broken bales to dispose of.
- **Allergies.** If a horse has a dust allergy you will not be able to use straw. It will be necessary to use paper bedding or shavings (however, some types of shavings can be dusty.) Rubber matting and paper are the best types of dust-free bedding.
- **Disposal.** Do you have someone who will buy the muck or will you have to pay for it to be removed?

Types of bedding	Advantages	Disadvantages
Wheat straw	Easy to obtain. Rots well to make compost. Fairly inexpensive. Horses tend not to eat it. Clean, bright appearance.	Prone to being dusty. Must be stored under cover. May be in short supply after a bad harvest. Heavy to handle when mucking out.
Barley straw	Makes a good, springy bed. Inexpensive. Rots well. Easy to obtain. Clean, bright appearance.	Horses tend to eat barley straw. Awns may irritate thin-coated horses.
Oat straw (Not really useful as bedding)	Can be used as bulk feed for ponies.	Horses love to eat it. Like all types of straw, if eaten, it can cause impaction colic.
Wood shavings	Non-edible. Good shavings are less dusty than straw. May be stored outside as bales are wrapped in polythene. Lighter than straw to work with.	Expensive. Can dry out the horse's hooves. Large bales are difficult to handle. Certain bales of shavings are very dusty, so cannot be relied upon as 'dust-free'. Cannot be used with a horse who has an open wound.
Shredded newspaper/ cardboard	Non-edible. May be stored outside as bales are wrapped in polythene. Lighter than straw to work with. Dust-free.	Can make a mess if allowed to blow around the yard. Expensive. If not well looked after, paper can become soggy and compacted, making it difficult to muck out. Difficult to dispose of.
Dust-extracted straw	As shavings – dust-free. Can be stored outside wrapped in polythene.	Expensive compared to ordinary straw.
Rubber matting	Saves labour as the floor is swept clean each morning as opposed to being mucked out. Saves on bedding costs. 100% dust-free. Non-slip. A small amount of shavings can be used to encourage the horse to stale.	As little or no bedding is used on the matting, the rugs become badly soiled. The matting is expensive to buy initially. Rubber floors without bedding do not look as aesthetically pleasing as a well-bedded stable. Some horses are not happy to stale or lie down on a rubber floor without bedding.

4.1 Bedding materials

ITQ 4.1

List five reasons why bedding is necessary:

1.

2.

3.

4.

5.

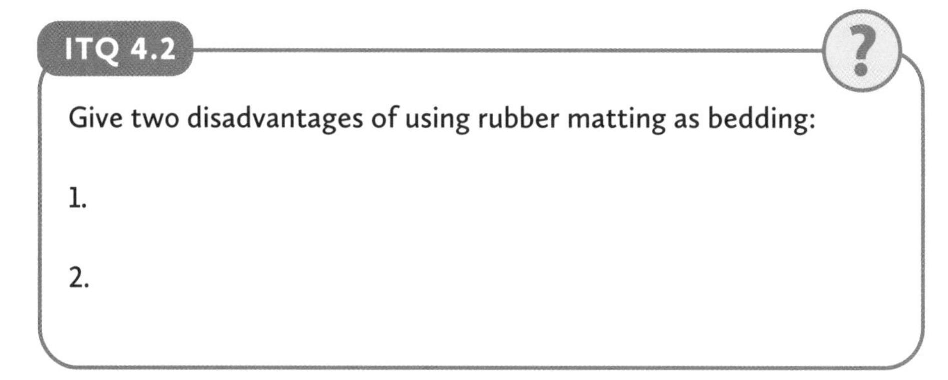

ITQ 4.2

Give two disadvantages of using rubber matting as bedding:

1.

2.

ITQ 4.3

List two types of bedding suitable for a horse with a dust allergy:

1.

2.

ITQ 4.4

Which type of bedding is not suitable for an allergic horse?

MUCKING OUT

Having looked at the types of bedding available, we now look at the methods of looking after the beds.

The stable must be mucked out every day. If the horse has been stabled overnight, this is normally done first thing in the morning. (For private owners, it may make sense to first exercise the horse, then turn him out and then muck out. That way, the horse isn't hanging around outside a possibly dusty stable while it's mucked out.)

Equipment needed

- **Wheelbarrow**. Used to cart muck to the heap. An alternative to a wheelbarrow is a **muck sack** – a large hessian or polythene square with handles in each corner. The main disadvantage of using a muck sack is that they are heavy to carry – you have to carry a full sack over your shoulder, which also means you end up smelling like the muck sack!
- **Four-pronged fork**. Used to remove the droppings and wet straw.
- **Shavings fork**. This is a specially designed, multi-pronged fork used to sift through dry shavings. The dung remains on the fork and the shavings fall through the prongs.
- **Broom.** After mucking out, always sweep the yard and leave everywhere very tidy. A wide broom should be used on large areas to save time.
- **Shovel.** Used to pick up the sweepings from the yard.
- **Skip.** This is usually a large plastic or rubber bowl and is very useful when skipping out a stable. The skip is also used when picking out the horse's feet.
- **Rubber gloves.** These can be worn to pick up the droppings when skipping out a stable.

Methods

There are two main methods of mucking out a stable:

1. The deep litter method
2. Mucking out completely

(If rubber matting alone is used, the process is simply to shovel up the muck and hose the floor.)

SAFETY TIPS

- Always use forks with great care. Forks can easily puncture a horse's leg or your wellington boot.
- Only one person should work in a stable with a fork. An accident may occur when two or more people work in a confined space, each using a fork.
- If you do use the fork when the horse is in the stable, take great care not to prod him **with it – always work with your back to him, directing the fork away from the horse.**
- Tools must be put back in their place after use. Tools left lying around are a danger to horses and humans. If forks are hung up, the prongs must point in towards the wall to prevent someone catching themselves on the pointed ends.

The deep litter method

This is the method whereby only the droppings and the worst of the wet bedding (usually straw with this method) is removed. It is not an ideal method for paper beds as they can become very soggy and compacted. Shavings beds can be kept on a deep-litter basis but it is not ideal to do so.

The advantages of this system are:

- It is labour-saving and quick on a daily basis.
- It provides a deep and warm bed.
- It saves on costs of bedding materials as less clean straw is needed.
- When mucking out there is less disturbance of the bed so less dust and fewer spores are released into the atmosphere.
- If properly managed it should not smell.

The disadvantages of this system are:

- The stable will need a proper mucking out once every six weeks or so, which is usually very hard work as so much heavy manure has accumulated.
- If not looked after properly, the bed may become soggy and smell of ammonia, which is not good for the horse's respiratory system. (Ammonia is excreted in the horse's urine and accumulates in the bedding, giving off pungent fumes.)

Procedure

- Gather your tools and put them neatly by the stable.
- Put on the headcollar and lead-rope, pick out the horse's hooves and bring him out of the stable and tie up safely. Alternatively, put the horse into a spare stable. If he has to stay in the same stable put a headcollar on and tie him up. This is not ideal as the horse is then in your way and the risk of an accident to both you and the horse is increased.
- Position the wheelbarrow at the door with the handles into the stable. This allows you to push the full wheelbarrow out of the stable easily. If the horse is in the stable, make sure he can't get caught in the barrow handles.
- Remove all of the droppings with the fork. Shake off all clean straw and save. Remove only the wettest patches of soiled bedding. Level the bed.
- Put in plenty of clean straw, level the bed and bank up the sides.
- Sweep and leave tidy.

Mucking out completely

This system can be used for straw, shavings and paper beds.

The advantages of this system are:

- As all wet bedding is removed daily there is no need for a massive muck out every few weeks.
- The floor may be swept, washed and disinfected frequently.

The disadvantages of this system are:

- It is much more time-consuming on a daily basis.
- More bedding material tends to be needed.

Procedure

- Remove the horse from the stable or tie up as for the deep litter method.
- Remove the droppings, making sure that you do not waste any clean bedding. Put all clean bedding into the corner. Remove all wet bedding and sweep the floor. Turn the banks over and clean and sweep under them.
- Load the wheelbarrow evenly, packing the muck down as you do so to stop it falling off or blowing around when you move the wheelbarrow.
- If very dirty, the floor will need to be hosed down. All stable floors should be thoroughly disinfected regularly (once a month or more).
- Leaving the floor to air and dry will freshen it, although it is not essential to do this every day.

- Spread the old bedding on the floor evenly. Put in a bale of clean straw or shavings/paper and spread this out, banking up the sides well. The bedding on the sides of the stable is banked up to help prevent the horse from becoming cast in the stable when he rolls and to help keep out draughts.

- Sweep the front of the stable floor and yard.

When the stable has been mucked out and bedded down with clean straw it is known as **'set fair'**. After mucking out the yard must be thoroughly swept and all tools put away.

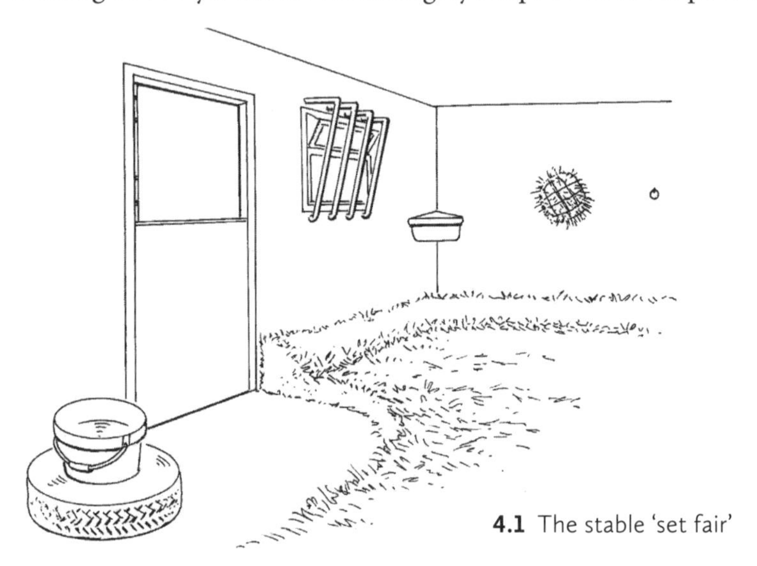

4.1 The stable 'set fair'

Skipping out

As well as normal mucking out in the morning, if the horse is in his stable for any length of time you should remove the droppings – this is known as skipping out. This should be done frequently to keep the stable clean. The most effective way to do this is to take a plastic skip into the stable with you and pick up the droppings by hand, using rubber gloves.

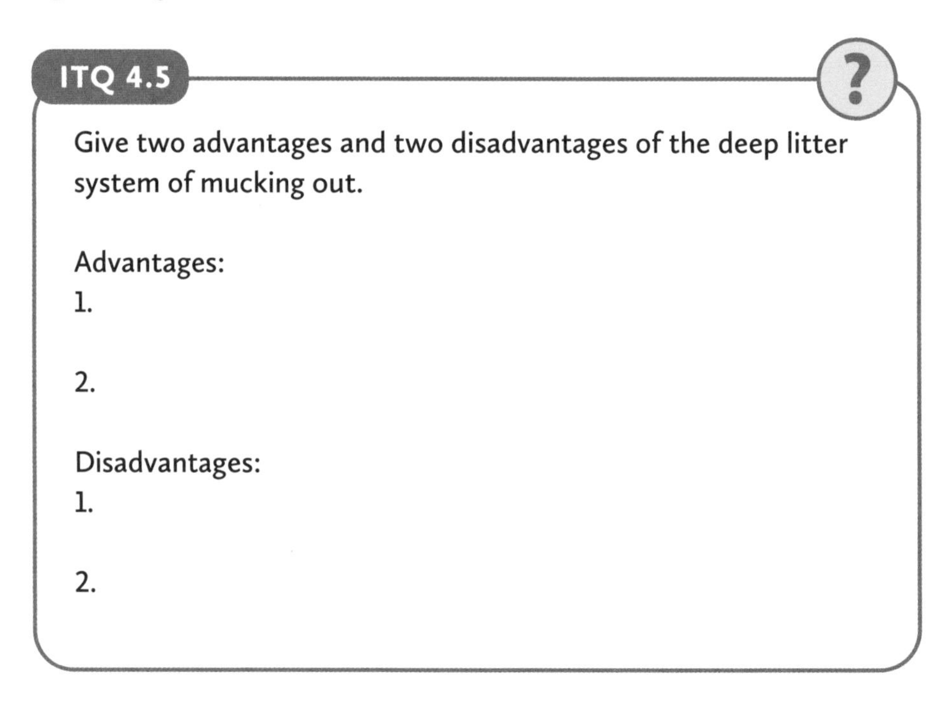

ITQ 4.5

Give two advantages and two disadvantages of the deep litter system of mucking out.

Advantages:
1.

2.

Disadvantages:
1.

2.

ITQ 4.6

Give two advantages and two disadvantages of the system of completely mucking out each day.

Advantages:
1.

2.

Disadvantages:
1.

2.

SAFETY TIP

- The horse should be tied up to prevent him walking about the box while you skip out. It is not necessary or practical to take the horse out of the stable every time you skip out.

The muck heap

The muck heap must:

- Be fairly close to the yard to save time when emptying wheelbarrows.
- Not be so close that flies and smell become a problem in the yard.
- Be away from drainage ditches and streams, to avoid pollution. Run-off from the muck heap must not be allowed to enter a water course.
- Have good access from the yard, on a firm, level and well-drained surface. It is much harder work to push a heavy barrow through mud.
- Have good access to allow muck collecting lorries and tractors to get to it.
- Be kept square and tidy. A three-sided bunker with a concrete base will contain the heap and prevent it from spreading. Every day, the heap should be packed down, squared off and the sides levelled. (Jumping on it is the best way to pack it down). Once packed down and squared, the sides need to be 'combed' with a fork to pull away loose straw and tidy it. Finally, the ground should be well swept around the heap.

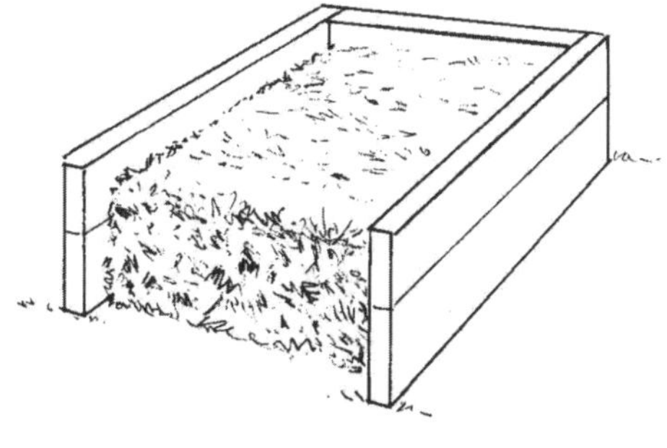

Tidy and square

Untidy

4.2 The muck heap

- Only ever be used for compostable waste, straw, shavings, shredded paper and manure.
- Never be used for the disposal of baler twine, tin cans, bottles, etc. Baler twine is a particular menace in the muck heap as it gets tangled in machinery and causes great damage.

Disposal of muck

It is usually possible to sell small amounts of well-rotted straw manure to gardeners. However, if there are more than two horses, the quantities of muck that need to be disposed of will probably exceed local demand and a manure collecting firm will need to be used, for which there is a charge. Paper bedding does not make such useful garden manure but it does break down fairly quickly.

A local farmer may take the muck to spread on the fields to improve soil quality. Mushroom growers use rotted horse manure – it may be possible to make an arrangement with a local factory. In urban areas, it may be necessary to hire a skip in which to empty the wheelbarrows. Manure used as fertiliser is not classified as controlled waste. If manure is not to be used as fertiliser it is classified as controlled waste and subject to Environmental Permitting Regulations – details can be found on the Environment Agency website.

EU regulations prohibit the burning of muck heaps.

5 Horse Husbandry – Describing Horses, Leading, Filling Haynets

REQUIRED SKILLS/KNOWLEDGE	Learnt, revised, practised?	Confirmed
Identify the points, colours and markings of a horse.		
• Describe horses' markings using the correct terminology.	☐	☐
• Describe coat colour using the correct terminology.	☐	☐
• Be able to indicate points of the horse.	☐	☐
Show that you can lead and hold a horse for treatment or inspection.		
• Correctly hold a horse for treatment or inspection.	☐	☐
• Lead a horse correctly in walk or trot.	☐	☐
• Turn the horse safely and correctly when leading in hand.	☐	☐
Correct use of haynets.		
• Fill and weigh a haynet.	☐	☐
• Tie up a haynet safely.	☐	☐
• Understand the potential dangers when using haynets.	☐	☐

DESCRIBING A HORSE

Height

The measurement of a horse's height is taken in **hands**: 'hh' stands for 'hands high'. There are 10cm (4in) to a hand. The horse is measured when standing on a level surface using a measuring stick. The measuring stick must have a spirit level on the cross bar. The horse must stand squarely with his head lowered so the eyebrows are in line with the withers. The cross bar of the stick is held across the highest point of the withers. If the horse is tense he should be allowed to relax before the final measurement is recorded.

If the horse is shod when being measured, 12mm (½in) is normally taken off the overall measurement. A pony is usually classed as being 14.2hh (147.3cm) or less. Shetland ponies are measured in centimetres or inches.

Accurate measurement is important because:

- It helps when selecting the correct size tack and equipment. The horse's build also affects sizing requirements.
- The correct measurement must be given when selling a horse.
- Horses and ponies are often divided and subdivided in competitions according to height.

Colour

When describing the colour of a horse, reference is sometimes made to the **points**. The points are the muzzle, tips of the ears, the mane, the tail and the extremities of all four legs.

Chestnut	A chestnut horse has a ginger or reddish coloured coat with similar points. A **liver chestnut** is a slightly darker version.
Bright bay	A chestnut body with black points.
Dark bay	Dark brown body with black points.
Black	Black body with black points.
Brown	Dark brown or nearly black body with brown points.
Grey	White coat and points. Horses destined to be grey will appear chestnut, black or dark bay at birth and will turn (increasingly light) grey as they age.
Iron grey	Fundamentally white coat with dark hair showing through.
Dappled grey	Iron grey with round darker grey dapples, usually on the hindquarters.
Flea-bitten grey	Grey, with reddish marks on the coat.
Cream	Cream-coloured coat. These horses sometimes have blue eyes (**wall eyes**).

Dun Varies from a cream colour to a dark golden body with black points. A black stripe known as an **eel stripe** runs down the length of the spine.

Roan A **strawberry roan** horse has a chestnut coat with white hairs showing through. A **blue roan** is a dark grey with either black, white or chestnut or a mixture each showing through. A **bay roan** is a bright bay with grey or white showing through.

Piebald Large irregular patches of black and white.

Skewbald Large irregular patches of white and any other colour except black.

(Piebalds and skewbalds are often referred to as 'coloured'.)

Albino Because of a genetic peculiarity, albino horses do not possess colour pigment. They have a body that appears light cream, and pink eyes.

Palomino Light cream to dark gold with similar coloured points. The mane and tail may be a lighter version, or silver.

Chestnut

Skewbald

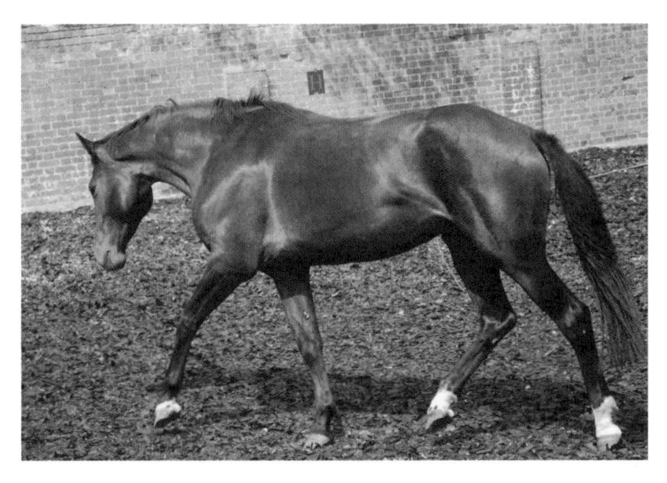

Dark bay

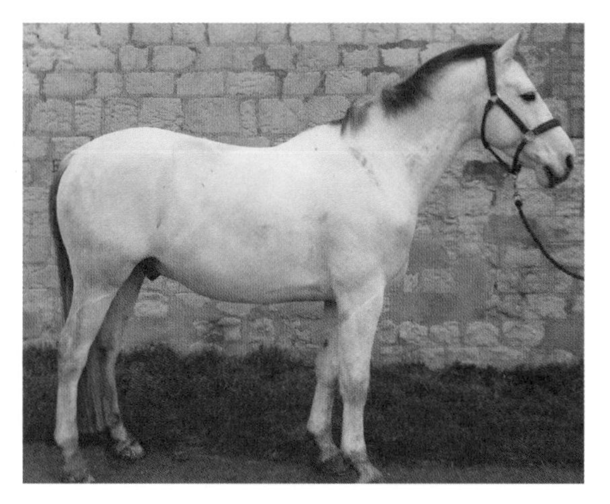

Grey

5.1 Horse colours

Markings

These (except whorls) are subsidiary, localised areas of colouration.

Star	Small white mark between the eyes.
Stripe	Thin white line running down the front of the face.
Blaze	Wide stripe down the face.
Snip	Small mark to the side of a nostril.
Ermine marks	Small dark marks on a white coronet or pastern.
Eel stripe	Dark line running down the length of the spine, normally seen on a dun.
Sock	White leg up to the fetlock.
Stocking	White leg up to the knee or hock.
Whorls	These are not areas of different colour, but patterns that form where the hair changes its direction of growth. As they vary between horses, they are recognised as identification marks.

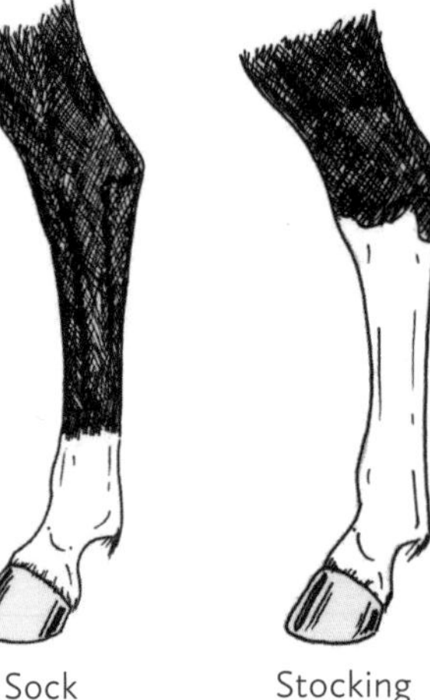

5.2 The markings

AGE

Animals of different ages are broadly defined in the following ways.

Foal In Thoroughbreds, a horse is a foal from birth until the following January 1: after this date, the foal becomes a **yearling**. In non-Thoroughbreds, a foal is a youngster up to the end of his eleventh month.

ITQ 5.1

a. What is the procedure for measuring a horse or pony accurately?

b. Give three reasons why it is important that horses are measured accurately:

1.

2.

3.

ITQ 5.2

What is the difference between a bright bay and a dark bay?

ITQ 5.3

What is a 'wall eye'?

ITQ 5.4

Whorls provide another means of identifying horses. What are whorls?

Filly	Young female up to 4 years of age.
Colt	Young uncastrated male up to 4 years of age.
Aged	Horse over 8 years of age.

Sex

The following definitions are used.

Gelding	A castrated male horse of any age.
Entire\stallion	Uncastrated male of four years or over.
Rig	A male horse with one or both testicles undescended (i.e. concealed in his abdomen). The correct term for a rig is a **cryptorchid**.
Mare	A female horse over 4 years of age.

Points of the horse

When describing horses the correct terminology should be used for the various anatomical points. It is well worth spending the time to learn these points.

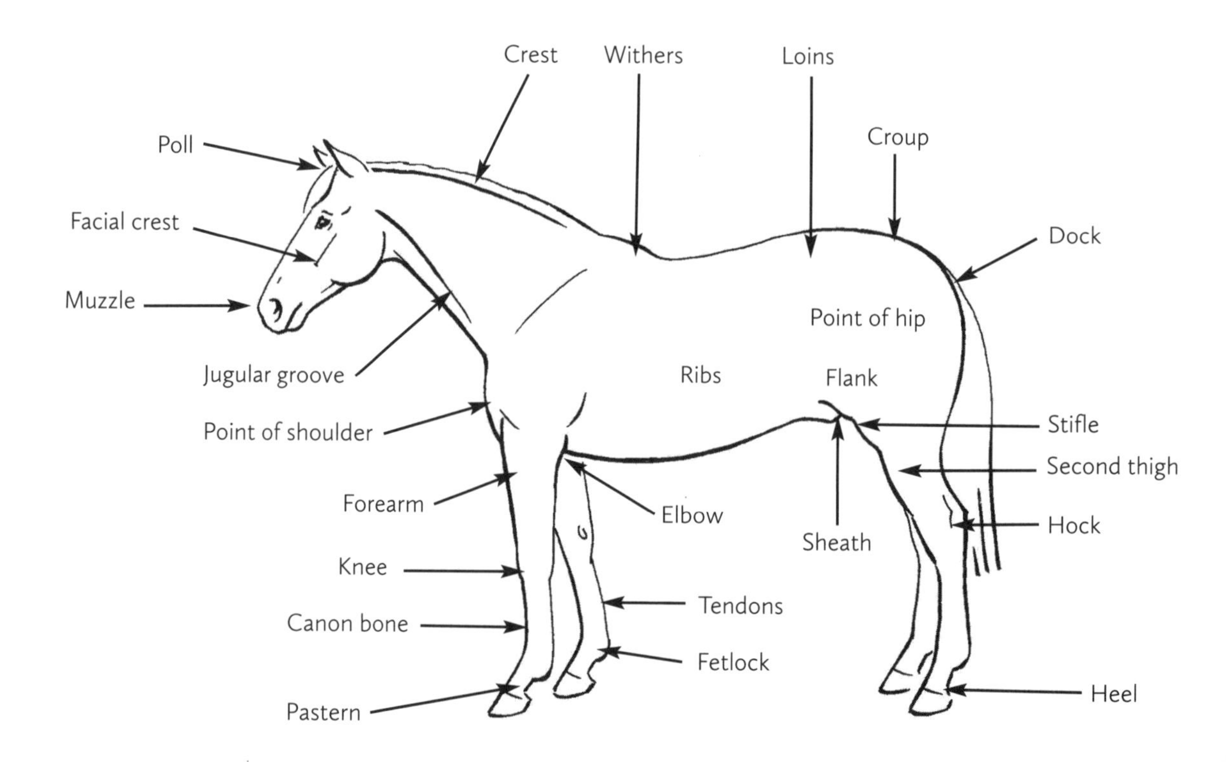

5.3 Points of the horse

LEADING AND STANDING UP

Clothing/equipment notes:

In the exam you must wear gloves and a crash cap when leading. At home it is safer to always wear gloves to prevent rope burns. If leading a young, unruly or unknown horse you should wear your crash cap at home too.

Carry a whip in the exam. When teaching a young horse to lead at home it is sensible to carry a whip but when leading under normal circumstances, e.g. to and from the field, a whip is not usually carried.

Leading from a headcollar

If leading from a headcollar, always use a lead-rope.

Generally it is not considered safe to lead from a headcollar only as, in the event of the horse pulling away suddenly, you would probably have to let go. If you did not let go, your hand and/or arm could be injured. When holding a rope, should the horse pull away, you have more chance of being able to hold on. It is safest to wear gloves when leading to prevent rope burns should he pull away.

Never wrap the rope around your hand or put your fingers through the metal/brass rings of a headcollar as this can cause serious injury if the horse bolts. Make sure the rope does not drag along the ground.

It is correct to lead from the horse's nearside, although he should be taught to be led from either side. Stand at the shoulder, give the command 'Walk on!' and walk beside him – never drag him along behind you.

Always look straight ahead, never look back at the horse – this often has the effect of making him stand and stare back at you. With practice the horse should become obedient to the voice, walking on and halting when asked.

To turn the horse, push him *away* from you so that you are on the outside of him as he turns. If you pull him *around* you there is a chance that he will tread on you or knock himself.

5.4 Leading the untacked horse

Leading in a bridle

If you are leading a horse you are unsure of, it is safer to use a bridle to lead in to give you more control. In an exam situation, if you are given the choice, use a bridle to lead the horse.

If leading along the road always put on a bridle for extra control. You should be on the left-hand side of the road and walk between the horse and the traffic, i.e. on the horse's offside (right). Wear a fluorescent tabard to make yourself more visible to traffic.

Leading a tacked-up horse

Take the reins forward over the horse's head. Standing on the horse's nearside, hold the reins in both hands with your right hand approximately 15cm (6in) from the horse's chin groove. Hold the whip in your left hand.

If the horse is wearing a running martingale:

- For short distances the reins are not taken over the head. Leave the buckle resting at the horse's withers – hold the reins in your right hand between the bit rings and the martingale rings.

- For longer distances or if you feel you need more control (e.g. if the horse is a bit lively) detach the martingale rings from the reins and secure to the neckstrap. Take the reins over the horse's head.
- Stirrups must be run up and the girth should be reasonably tight to stop the saddle slipping.
- Whether at walk or trot the horse must be moving actively forward on a straight line.

Leading in an exam

- Make sure the yard gate is closed and no unsafe items, e.g. wheelbarrows, tools etc., are in the way.
- Wear gloves, a fastened crash cap, and carry a whip in your left hand.
- Lead the horse from the nearside unless instructed otherwise.
- The horse should always have a bridle on for control.

Trotting up

- When trotted up, the horse must be active (but not excessively speedy).
- If he is dragging behind, you must keep looking forwards and encourage him with your voice and, if that is not enough, use a flick of the whip behind you, lightly touching the horse's side.
- Be prepared for the fact that some horses may become over-enthusiastic and try to trot too quickly or pull away while trotting up; use your voice and pressure on the reins to steady the horse.
- Allow enough time to move the horse from halt, through walk to trot and back again. When trotting back towards the examiners make sure you allow enough time to bring the horse smoothly to a halt.

Standing a horse up for inspection

When standing a horse up for inspection you should stand in front of the horse on a level surface. The reins should be forward over the horse's head, with one rein held in each hand close to the horse's head. Hold the slack of the reins and the whip in the right hand. If you feel the horse may barge forwards you can stand slightly to one side.

If the horse is wearing a headcollar, do not hold him with your hand through the headcollar. This could make a sensitive horse pull back, which could be dangerous. Always hold the lead-rope.

The horse should stand square, i.e. his weight should be evenly distributed over each limb. The horse must not be allowed to rest a hind leg whilst standing up for inspection.

5.5 Standing the horse up for inspection

SAFETY TIP

- You should always stand on the same side as the person treating the horse as he is likely to try to move away from them.

Holding a horse for treatment

If you are holding a horse for treatment it is safer to have him in a stable to give extra control. Position the horse with his hindquarters towards the corner of the stable to help stop him moving about too much.

If the horse needs to be restrained you can hold up one flexed foreleg, holding it at the cannon bone. Keep your back straight and hold it firmly. If you need to let go, tell the person administering the treatment.

Another simple restraint technique involves flexing the horse towards you and firmly grasping a large handful of skin on the neck.

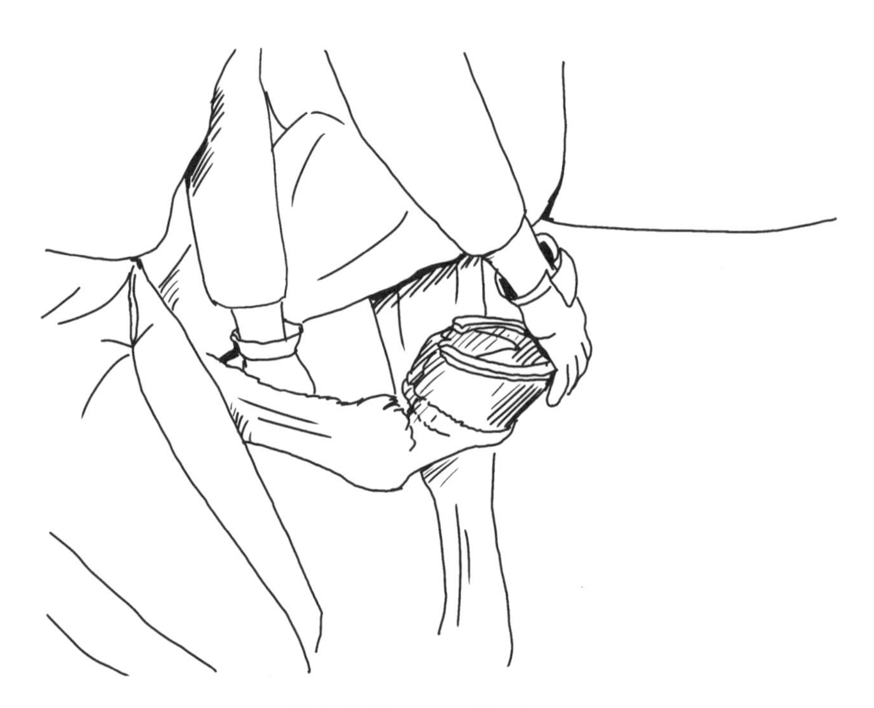

5.6 Holding up a foreleg

FILLING AND HANGING HAYNETS

One or two hooks placed approximately 1m (3ft 3in) off the ground on the wall in the haystore makes filling a net easier. Hook the empty net onto the hooks. Put the hay into the net in whole sections and shake up lightly once in the net.

Once full, hang on a spring weighing hook to ensure you have the correct quantity of hay in the net.

The haynet must be hung from a ring (not using a weak link) which is securely fastened to a solid object such as a post or wall.

Pull the haynet strings to close the net and ensure the knot joining the strings is at the very end. Pass this end through the ring and pull the net up until it reaches the ring.

Pass the strings through the haynet approximately three-quarters of the way down the net and pull the strings upwards towards the ring. Fasten the strings using a quick-release knot and pass the end of the strings through the loop. Twist the haynet so that the quick-release knot is at the back, away from the horse.

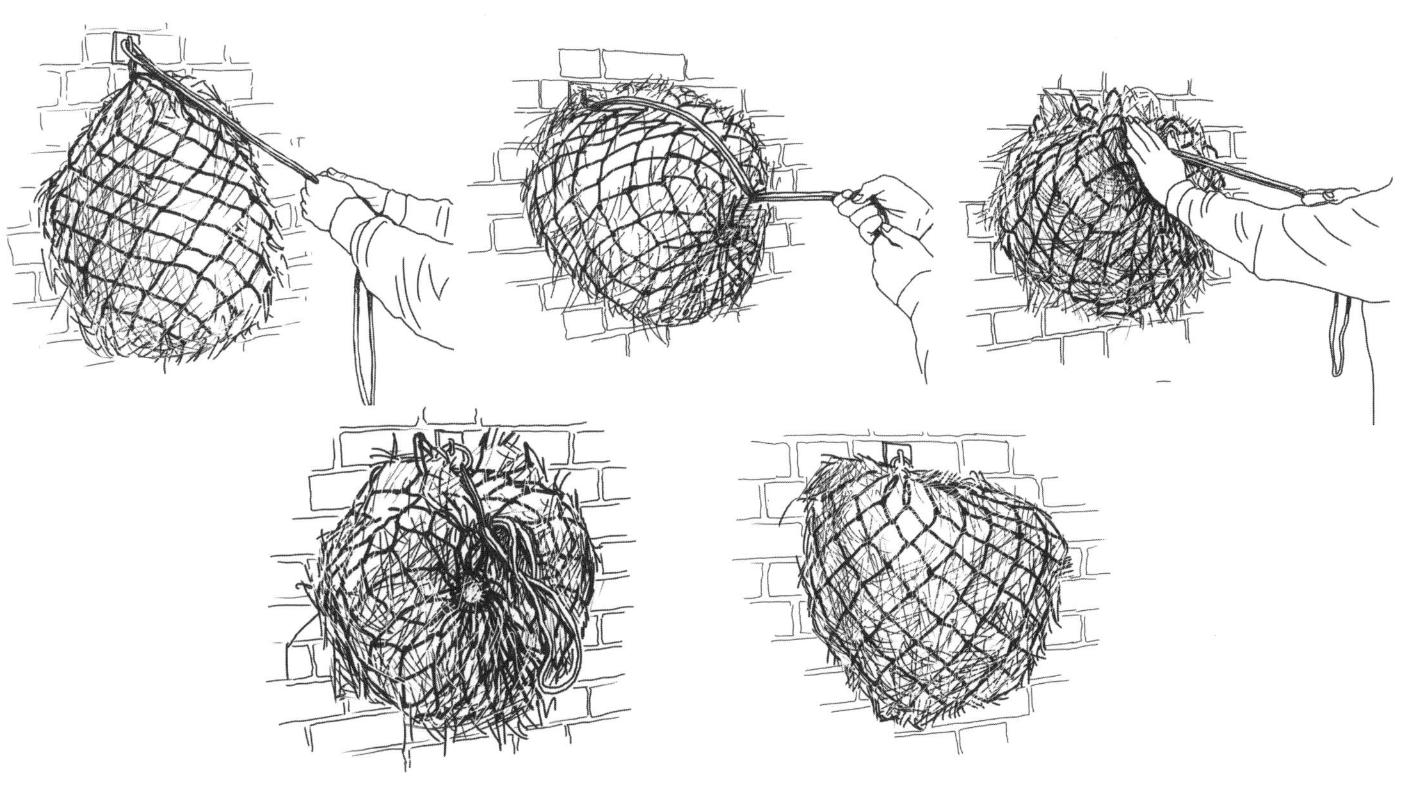

5.7 Tying up a haynet

SAFETY TIP

- When empty, a haynet will hang lower than when full. Make sure it is tied high enough when full to prevent it hanging low enough for a horse to get a foot caught when empty.

The haynet must not be hung above the horse's eye level, or hay seeds will fall into the horse's eyes. If hung too low the horse could get a foot caught in it.

Never use haynets when feeding foals and try to avoid using them in fields – it is safer to feed hay from the ground in the field.

If the haynet has been soaked in water to minimise dust it will be very heavy when you first remove it from the water. Allow it to drain before moving it.

ITQ 5.5 ?

List two dangers of an incorrectly tied haynet:

1.

2.

6 Clothing, Health, Fire Precautions, Accident Procedure, Riding Out, The BHS

REQUIRED SKILLS/KNOWLEDGE	Learnt, revised, practised?	Confirmed
Understand the basic principles of health, safety and welfare when working with horses.		
• Describe suitable clothes to wear when working with horses.	☐	☐
• Understand the importance of physical fitness when working with horses.	☐	☐
• Fire precautions.	☐	☐
• Basic accident procedure.	☐	☐
• Rules and protocol for taking horses out on a public highway.	☐	☐
• The role of the BHS in promoting horse and rider safety and welfare.	☐	☐

CLOTHES FOR WORKING WITH HORSES

When working with horses you should appear professional, workmanlike and practical, looking as neat and tidy as possible. Each yard will have its own dress code.

Working with horses involves being outside in all weathers including everything from freezing wind, rain and snow to heatwaves; it is important to have the right clothing to be able to work effectively. Outdoor pursuit shops sell a huge range of technical outdoor clothing including base layers, gloves, hats, fleeces and jackets, which are all ideal for working outdoors.

Clothing worn when working with horses needs to be practical, keep you at a comfortable temperature and offer a level of protection from cold, heat, tread injuries and rope burns.

Base layers. Depending on the weather, you should aim to use layers of breathable clothing so you can adjust the amount you wear according to your body temperature.

In hot weather it looks more professional to wear short-sleeved polo shirts rather than skimpy sleeveless vest tops (and helps stop you getting sunburnt shoulders).

Jacket. You will need a decent waterproof jacket, preferably one with breathable qualities so as you warm up it wicks perspiration away, keeping you dry. A decent hood is essential.

Overalls. If you are mucking out several stables, overalls can protect your clothes, keeping them clean and stopping them from smelling.

Waterproof overtrousers. In torrential rain, waterproof overtrousers are needed. Again, these should ideally be of a breathable material with the lower section zipped so they can be pulled on and off over wellington boots. It is possible to buy waterproof overtrousers suitable for riding.

Footwear. Boots or shoes must always be sturdy, secure and non-slip, even in very hot weather. Clogs, flip-flops and soft shoes are unsuitable. It is not essential to wear steel toe-cap boots as these can cause further injury if a horse treads on the steel rim. Wellington boots are needed for walking in muddy, wet fields and for mucking out. Don't wear leather boots for mucking out as the urine rots the stitching and damages the leather.

Socks. When riding or standing (e.g. if teaching or lungeing) in very cold weather your feet can get extremely cold, to the point of being painful. It is worth investing in decent hiking socks, including thermal inner socks, to keep your feet warm. Wearing several pairs of socks can restrict your feet and make them feel even colder.

Gloves. Non-slip gloves are needed all year round for leading and lungeing horses. When working in cold weather you will need warm, preferably waterproof gloves. If trying to preserve your fingernails and prevent calluses, gloves should be worn for stable duties, especially mucking out.

Hat. In very cold weather you will need a warm winter hat. In hot, sunny weather a baseball cap can help keep the sun off your face.

Crash cap. A correctly fitted crash cap should be worn when riding or lungeing. It should also be worn when leading and handling youngsters and unruly horses.

Hair ties. Long hair should be tied back when working with horses or riding.

Jewellery. Ideally no jewellery should be worn but if it is, it must be kept to a minimum. Loop earrings, necklaces and studded rings, etc. can be dangerous as they can get caught. Rings can scratch saddles.

Perfume. Strong perfumes should not be worn as horses are very sensitive to smell.

PHYSICAL WELL-BEING

Working with horses is physically demanding and a good level of general fitness and strength is necessary. Your fitness will develop the more you work with horses but the risk of injury increases for those who are not particularly fit or who suffer from health problems such as a bad back or arthritis. Anyone wishing to work with horses who has a long-standing health issue or physical impairment is advised to seek medical advice and to inform prospective employers of the nature of the condition and of any particular procedures or medication that may need to be applied in the event of a health problem or accident.

FIRE PRECAUTIONS

The Regulatory Reform (Fire Safety) order 2005 applies to all private and commercial stables and equine establishments and imposes responsibilities on the 'responsible person' of the premises to make a 'suitable and sufficient' assessment of the risks to identify the general fire precautions they need to make to comply with the requirements and prohibitions imposed on him by or under this order.

Because of the large amount of hay, straw and shavings stored, and the fact that many stables are of wooden construction, stable yards are very susceptible to fire.

Safety rules

There are certain rules that must be enforced on the yard to help prevent fire. These include:

- Enforce a strict NO SMOKING rule and display signs to that effect.
- Make sure that electrical wires are well maintained. Mice can chew the outer casing, which can lead to problems. All electrical appliances must be kept in good working order and checked regularly by an electrician.
- Use a residual current device when using electrical items and make sure that all trip switches on the main fuseboard are in good working order. This ensures that if a fault occurs in an electrical item, the power supply is instantly cut off.
- Take great care with electric or paraffin heaters – never leave them on unattended and make sure they are not positioned near chairs, etc.
- Clean dust and cobwebs away from all light bulbs and ensure that lights are covered with a wire mesh if there is a chance that the horse can reach the bulb.
- Always stack new hay quite loosely, allowing space for air to circulate. This is because new hay, especially if not completely dry, can heat up in the stack and in

extreme circumstances may combust spontaneously.

- Flammable materials such as gas bottles, petrol cans, paint, etc. should be stored in a separate building away from the stables.
- Never light bonfires or hold firework displays near the yard. As well as frightening the horses, stray fireworks are a common cause of stable fires.
- Never use welding equipment near anything flammable such as hay or straw.

The fire point

Every yard should have at least one fire point – in a commercial yard it is a legal requirement. The fire point should be in a clearly visible position and consist of the following:

Notice displaying the action to be taken in the event of fire. This should give the following information:

- The location of the nearest telephone.
- The address of the yard.
- Instructions to dial 999 and request the Fire Brigade.
- Instructions to sound alarm – ring bell.
- Instructions to move horses in greatest danger first, using headcollars if time permits.
  - If there is no time – open doors to release horses. (**NB** horses may be frightened and unwilling to leave their stable. Blindfolding with a coat or similar may encourage them to walk out of the stable.)
  - If possible put the horses into a safe paddock or school.
  - Shut both doors of empty stables to prevent horses returning.
- Instructions to tackle fire without endangering life.

Fire extinguishers suitable for dealing with different sorts of fire. Ask the local Fire Prevention Officer for advice regarding the type and number of appliances needed. **NB** fire extinguishers must be serviced annually and everyone on the yard must know how to use them.

Two water-filled buckets and two sand-filled buckets.

A full water tank – salt may need to be added to prevent it from freezing in winter.

A hosepipe attached to the tap (This must have a good pressure and be capable of reaching the furthest stables in the yard).

An axe.

Everyone must be briefed on the fire drill and it should be practised in large yards.

BASIC ACCIDENT PROCEDURE

This section is intended to introduce basic accident procedure. Everyone working with horses should train with a recognised organisation and gain the First Aid at Work Certificate. Refresher courses need to be undertaken at regular intervals to maintain skills and keep up to date with new developments. The British Horse Society organises Equine Specific First Aid Courses which, as the name suggests, cover the type of accidents which affect those handling and riding horses.

Action to be taken after a fall or accident

1. The ride should halt somewhere safe.

2. Depending on the circumstances, the most experienced rider or the instructor should assume control unless one of the other riders is medically qualified.

3. Assess whether there is danger to you – it is important that you do not become another casualty.

4. Assess the situation so that suitable control procedures can be delegated. For example, if the accident occurred on the road, someone should be positioned on either side of the accident to slow down and control the traffic in front and behind. If the horse is loose, someone competent should be sent to catch him.

   At this point, your assessment of the situation will indicate whether the casualty needs to be moved or not. For example, an injured person lying in an icy, water-filled ditch on a freezing day is in danger of drowning or becoming hypothermic so needs to be moved.

   Because of the risk of spinal injury, never move the casualty unless it is absolutely necessary. Incorrectly moving a casualty with a spinal injury can result in permanent paralysis.

5. When the safety measures above have been taken to prevent the situation becoming worse, you can turn your attention to the injured rider. You must establish whether the casualty is conscious or not and provide emergency care as necessary.

   Do not remove the casualty's crash cap unless it is essential.

SAFETY TIP 

- Specialist life-saving techniques must be learnt so you are able to cope in an emergency situation. Everyone involved with horses should attend a course to learn these.

Remember the letters **DR ABC**:

D – Danger – check that you and the casualty are not in danger.

R – Response – try to get a response by asking questions and gently shaking their shoulders.

A – Airway – the airway should be clear and kept open. Place one hand on the forehead, two fingers under the chin and gently tilt the head back.

B – Breathing – normal breathing should be established. Once the airway is open check breathing for up to 10 seconds by looking for the rise and fall of the chest, feeling for their breath on your cheek and listening.

C – Compressions – if the casualty is not breathing you should call for an ambulance and start cardio-pulmonary resuscitation (CPR) (also known as mouth-to-mouth resuscitation) straight away. CPR is a combination of rescue breaths and chest compressions to keep blood and oxygen circulating in the body.

CPR method for adults

- Place your hands on the centre of the casualty's chest and, with the heel of your hand, press down 4–5cm (approx. 2in) at a steady rate, slightly faster than one compression a second.
- After every thirty chest compressions, give two breaths.
- Pinch the casualty's nose. Seal your mouth over their mouth and blow steadily and firmly into their mouth. Check that their chest rises. Give two rescue breaths, each over 1 second.
- Continue with cycles of thirty chest compressions and two rescue breaths until they begin to recover or emergency help arrives.

CPR method for children

- Open the airway by placing one hand on the forehead and gently tilting their head back and lifting the chin. Remove any visible obstructions from the mouth and nose.
- Pinch their nose. Seal your mouth over their mouth and blow steadily and firmly into their mouth, checking that their chest rises. Give five initial rescue breaths.
- Place your hands on the centre of their chest and, with the heel of your hand, press down one-third of the depth of the chest using one or two hands.
- After every thirty chest compressions (at a steady rate, slightly faster than one compression a second) give two breaths.
- Continue with cycles of thirty chest compressions and two rescue breaths until they begin to recover or emergency help arrives.

The recovery position

A casualty who is breathing but unconscious should be placed into the recovery position to ensure the airway remains clear and open and prevent vomit or fluid from causing choking.

- Place the casualty on their side so they are supported by one leg and one arm.
- Open their airway by tilting the head back and lifting the chin.
- Monitor breathing and pulse continuously.
- If injuries allow, turn the casualty onto their other side after 30 minutes.

If you think a spinal injury may have been sustained, do not move the casualty; place your hands on either side of their face and gently lift their jaw with your fingertips to open the airway. Take care not to move the neck. If breathing is or becomes noisy then place the casualty in the recovery position.

If the casualty is conscious:

- Offer reassurance and tell them not to move until their injuries have been assessed.
- Ask them if there is any pain and, if so, where.
- Ask them whether they can move their fingers and toes. If they are unable to do so, there is a strong possibility of damage to the spine, so the casualty *must not be moved* until skilled help arrives. Loosen clothing around the neck and waist and cover the casualty with a jacket to keep them warm.

Further procedure

If necessary, send someone to call an ambulance. Make sure the person calling the ambulance knows the exact address/location of the accident. Searching for a telephone in a rural area can waste time – this is where the mobile phone is so useful. Always keep the phone battery charged for this reason and take it with you on hacks.

Meanwhile, continue to reassure the casualty and check bleeding, which should be stemmed (see below) unless pressure to the wound would make matters worse e.g. pushing foreign matter further into the wound. Immobilise fractures as best you can with the equipment you have with you. Jumpers can be made into slings and clean handkerchiefs make useful pressure pads to stem bleeding. The casualty should not be offered anything to drink in case surgery is necessary, for which the stomach needs to be empty.

Controlling bleeding

Heavy blood loss leads to a reduction in blood pressure, which causes life-threatening shock. To control bleeding:

1. Apply direct pressure with the fingers to the bleeding points. If the wound area is large, press the sides of the wound together firmly but gently.
2. If the wound is on a limb, raise the injured part and support it in position. *Don't* do this if you suspect an underlying fracture.

3. Do not try to remove foreign bodies from a wound as this could cause further damage to veins or arteries.

4. When you have a clean dressing available apply it to the wound and press down gently but firmly. Cover with a pad of soft material and bandage in position.

Take the casualty to the nearest Accident and Emergency Department.

How to continue

If the fallen rider gets up straight away and doesn't suffer dizziness, double vision or complain of a headache, and has no obvious injury (other than to their pride!), they may be permitted to remount and continue the ride.

However, if there is any doubt as to their well-being, someone from the stables should be contacted to come and collect them by car and the horse should be led home. A rider who has been unconscious, even for a short time, must not be allowed to remount and should be taken to hospital for a check-up. There is a risk of delayed concussion in these cases, so such a person should not drive themselves to hospital.

The injured horse

If the horse is injured it may be possible to lead him home, or transport may need to be arranged. A vet may need to be called to meet the horse back at the stables or, in more severe situations, the vet may need to attend the scene of the accident. Try to keep the horse calm and, in cold weather, keep him warm with jackets or blankets. It may be necessary to keep one of the other horses beside the injured horse while waiting for help to stop him fretting about being separated.

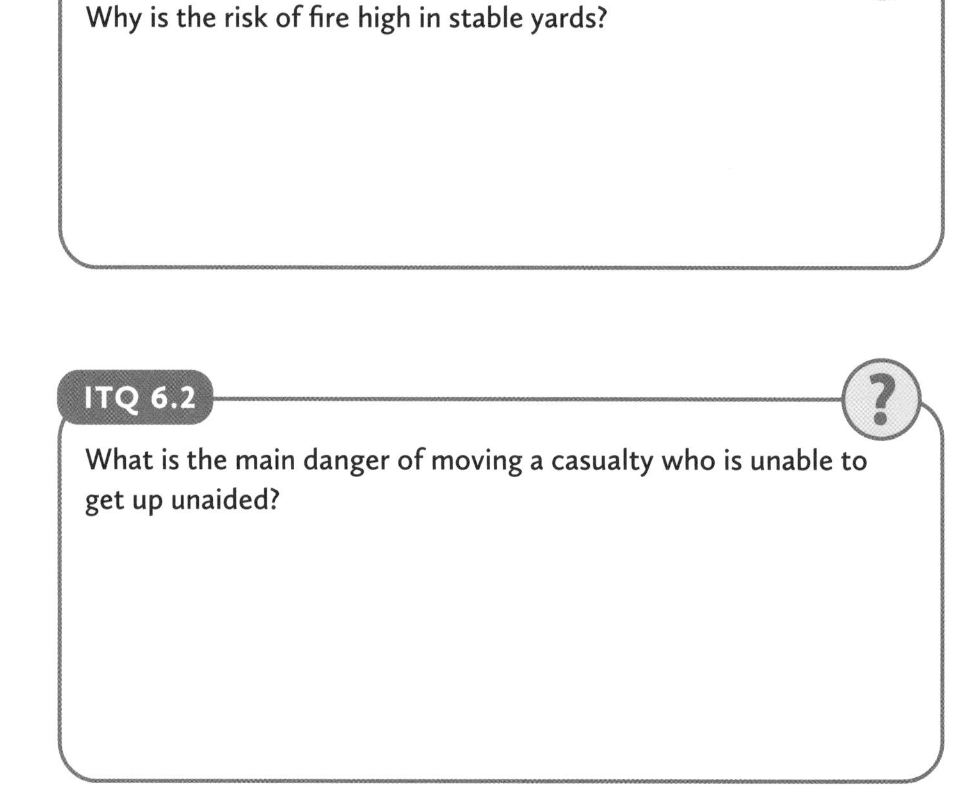

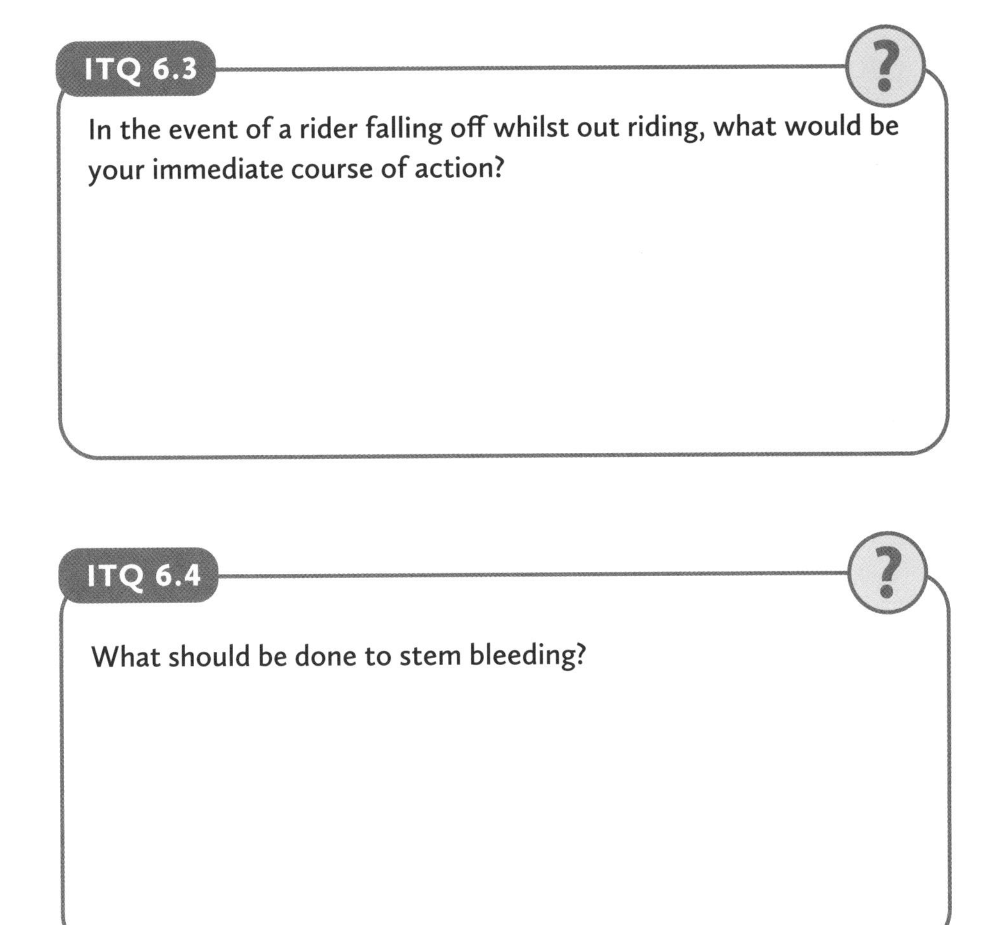

The accident report book

All riding schools and livery yards should keep an *Accident Report Book* to ensure all details of incidents are recorded in case claims are made against the insurance. All falls and any accident, whether it occurred whilst riding out or in the yard, should be recorded in the *Accident Report Book*, no matter how minor the incident may seem at the time.

This is best done whilst everything is still fresh in your mind and while witnesses are available to sign the book to confirm your description of events.

The information to be recorded in the *Accident Report Book* includes:

- Date, location and time of the accident.
- A diagram depicting the location of those involved and how the situation progressed.
- Names and addresses of those involved, including witnesses.
- Description of events, i.e. what happened and why.
- Record of injuries sustained and to whom.
- Record of any treatment given and by whom.
- Details of admission to hospital if that was necessary.

- Signatures, preferably, of all parties involved including the escort or instructor in charge at the time.

Advising insurers

The relevant insurance company should be advised if there is injury to a horse or rider, or damage to property, a motor vehicle or similar which could result in a claim. If in doubt it is always better to err on the side of caution and advise the insurance company.

Learn the sequence of priorities in the event of an accident so that if you are ever involved in one you will know what to do, and will not waste time panicking.

The golden rules if you are in an accident involving other people are:

- Never admit liability.
- Never apologise.
- Never accept the blame.
- Exchange names and addresses with the people involved in the accident.
- Ask for the name and address of their insurance company.
- Take the names and addresses of all witnesses.

If the police are involved they will also take everyone's details, but you will need to know them for your insurance company, as well as to fill in the *Accident Report Book*.

RIDING ON PUBLIC AND PRIVATE LAND

Many counties have large areas of common land and Forestry Commission land which horse riders are allowed to use. If you are lucky enough to have access to these areas do make sure you know and abide by any by-laws and regulations. It is also important that you keep to marked paths and are courteous and considerate to other people using the area. It is irresponsible to abuse this privilege as your behaviour may result in access to these areas being withdrawn.

It is not a criminal offence to ride on a public footpath unless there is a traffic regulation order or relevant bylaw prohibiting equestrian use. If you ride on a public footpath you will be trespassing unless you have the permission of the landowner to do so. However, if there are unrecorded rights on the path, and this were proven, it would not constitute trespass.

Whether you are riding on private land or a bridleway always:

- Keep to the edge of crops; do not ride through the middle of them.
- Keep off the fields if they are very wet, or have been re-seeded.
- Walk (as opposed to trotting or cantering) through fields containing livestock.

- Avoid fields with livestock in them if you are riding and leading as it can become an uncontrollable situation, especially if the livestock are inquisitive.
- Leave all gates as you find them.
- Dismount to close gates in fields containing livestock as it makes it easier to prevent them from escaping.

Bridleways

Bridleways are a public right of way to pedestrians, cyclists and drivers of horse-drawn vehicles who must give way to riders. If you wish to go past another rider or user of the path always ask their permission first. Ride past at walk to avoid upsetting their horse or splashing them if the ground is wet. Remember that many pedestrians are frightened of horses, especially when they pass too close to them.

If a bridleway is blocked, or if the landowner is offensive, do not enter into discussion but return the way you have come, and report the incident to your local British Horse Society Bridleways Officer. The British Horse Society has an Access and Rights of Way department which aims to keep bridleways, byways and permissive routes open and safe to use. Each region has its own Access officer to whom you should report any problems that you encounter. Places to ride can be found at www.emagin.org

Basic road safety

If visibility is poor it is safest not go out on the roads. Wear a fluorescent/reflective jacket or tabard so motorists can see you when hacking, especially if there is a risk of failing light, fog or generally murky conditions. Horses can wear reflective high-visibility leg bands. Stirrup lights that shine white to the front and red to the back can also be fitted. However, it is safer to avoid these conditions. It is also safer to avoid busy roads, built-up areas and known hazards such as heavy work machinery which can spook horses. In the modern countryside, this is less and less easy to do.

If hacking out in a group, everyone must remember that they have to obey the rules of the road like any other road user (see later this section). Stay on the left-hand side. Riders may ride two abreast on any road in a defensive manner.

The leading escort should make constant checks of the ride and should always be aware of the traffic in front and to the rear. When the road ahead is clear, the escort can wave traffic past that has been held up by the ride. The rider has a better view of the road from the horse's back than the motorist.

Traffic should never be allowed to divide the group up. The group should be maintained as a single unit, so when one horse turns across the road, the whole ride follows in close order. To facilitate cohesion, safety and minimal inconvenience to other road users, groups should not normally exceed eight in number, with a minimum of half a horse's length between each horse. In the event of a large number of riders, such as a sponsored ride, needing to use the road, they should leave a space of at least 30m (100ft) between each group so cars can overtake each group separately.

Courtesy to all drivers should become second nature – a smile and a nod cost nothing but may encourage drivers to slow down next time they pass a horse and rider on the road.

Hack young or traffic-shy horses out with a sensible companion to set them a good example and instil confidence.

Beginner riders should never be allowed to hack out. All riders should be competent at walk, trot and canter before going out and should be suitably mounted on sensible horses. Accidents can happen whilst out hacking as the horses are usually a lot livelier than when in the school. This can take the rider by surprise, especially if inexperienced.

Stud nails can be used in the shoes to give the horse extra grip when hacking on the road. Don't allow or encourage horses to hurry down hill on poor road surfaces.

Preparation and safety procedures for riding on the road

- Make sure you can control the horse before taking him out on the roads. If your horse is nervous, ride with a companion who is not nervous.
- Tell someone at the yard your intended route and have money or a card for the telephone. It is preferable to have a fully charged mobile phone with you.
- Always wear an approved safety hat/crash cap with the chin strap fastened.
- Wear a fluorescent/reflective tabard, regardless of the weather conditions. Wear something bright and light. If you must ride in the dark, in addition to fluorescent/reflective clothing, a stirrup light with a white light to the front and a red light to the rear on the outside leg is the minimum lighting that would be acceptable.
- Only wear boots or shoes with hard, smooth soles and a well-defined heel. Plimsolls, trainers, wellington boots and muckers are not suitable.
- Tack must fit correctly and be in good condition. You must never ride without a saddle or bridle on the public highway.
- Do not carry another person on the horse.
- Do not carry anything that might become entangled in the reins, or affect your balance and control. Make sure exercise sheets and rain sheets are properly secured.
- Keep both hands on the reins unless making hand signals.
- Both feet must be kept in the stirrups unless you are going over ice or an otherwise very slippery surface (both of which should be avoided if possible).

Adverse weather conditions

Fog and mist. Do not hack out in these conditions – give the horse a day off, or ride in the school.

Dusk or darkness. Avoid riding in these conditions if possible. If you *have to* ride out, wear reflective clothing and carry lights.

Slippery surfaces. Corners, drain covers, steep cambers, worn patches on the road, plastic paint marking white lines and zebra crossings are slippery. If such areas cannot be avoided, cross them carefully at walk – on no account attempt to cross them at a faster gait.

Snow and ice. It is not safe to ride out on the roads in icy conditions so these should be avoided. If it is absolutely essential to ride out, for example on private land or a driveway on the way to the school, keep to a walk and let the horse take his time. Before leaving the yard pack the horse's feet with grease to prevent snow compacting in the hooves. Keep to any grittier parts of the track and take your feet out of the stirrups and cross the stirrups over.

If the horse does slip over, let him regain his balance and check for injury. Road studs can improve grip but it must be reiterated that these conditions should be avoided whenever possible.

Windy weather. Horses tend to be very lively and/or spooky in windy weather. Be aware that you may not be able to hear approaching traffic very well. It is not safe to ride out in gale-force winds.

Low, bright winter sun. This may sound an unlikely 'adverse' weather condition but if you have driven a car on a bright winter's morning when the sun is low, you may have experienced its blinding effect. This is especially dangerous when driving alongside a hedge or wall which throws the road into a shadow. Riders (or pedestrians and cyclists) cannot be seen in these conditions.

If you are unfortunate enough to have an accident you should report it on www.horseaccidents.org.uk

Courtesy and common sense

Always acknowledge the consideration of other road users. Thank other road users who have slowed down or given you the right of way. This may encourage drivers to slow down for the next horse and rider they meet. Likewise, a lack of courtesy may deter a driver from slowing down in future.

If both hands are required on the reins for control, smile and give a nod of the head. If it's safe to do so, take one hand off the rein, raise your hand and smile to thank them. If their window is open, say 'Thank you'.

Be considerate and helpful to other road users. Request permission from other riders or pedestrians before passing them on the road (or on bridlepaths and tracks) and always pass at walk. Never trot or canter up behind other horse riders – this may upset their horses and cause an accident.

Avoid upsetting the horse in traffic; the public highway is not the place to discipline him or to teach him manners. Ride straight and control the horse's quarters with your leg aids and use of a whip if necessary.

Only dismount if absolutely necessary. You are usually safer on the horse's back. Dismount from either side. Keep the horse facing in the direction of the moving traffic. Remount using a verge or gateway and keep the horse facing in the direction of the moving traffic.

Look, listen and think ahead. Be aware and stay alive.

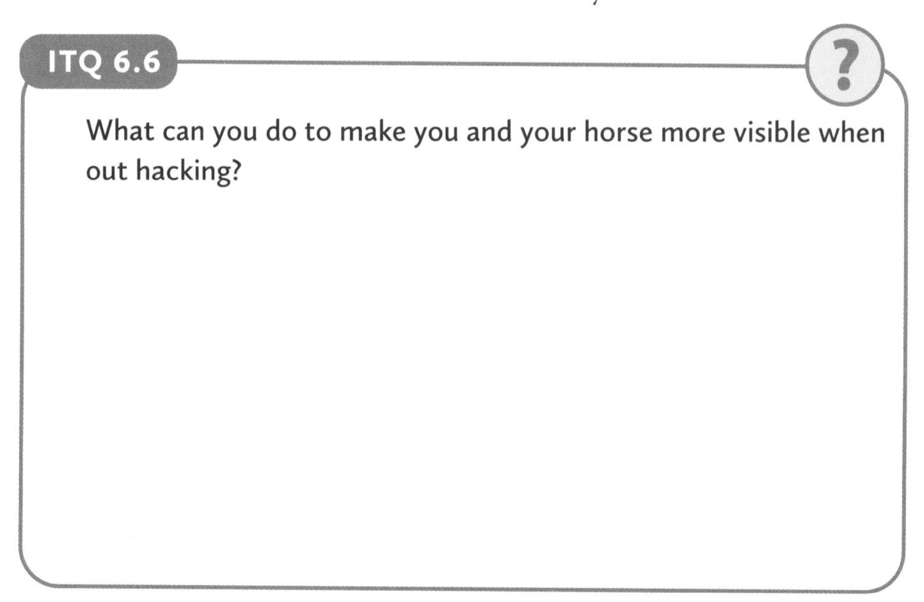

The Highway Code and riding code as applicable to those in charge of horses

At some time during your riding career you will have to negotiate public roads. Many riders have to use the roads for a vast majority of the horse's exercise. It is therefore very important that you know the *Highway Code* as it applies to horses and other road users.

It is highly recommended that all riders take the BHS Riding and Road Safety Test. The British Horse Society has published a book called *Riding and Roadcraft* which would be a very useful addition to your library.

Statutory rules in the Highway Code affecting those in charge of horses

You must not ride, lead or drive a horse on a pedestrian footpath (pavement) in England and Wales, or on a footway or footpath, unless there is a right to do so, in Scotland. Always use a bridlepath where possible. Also, you should not take a horse onto a cycle track. Only ever go onto the pavement in an emergency. Do not ride on verges belonging to houses.

Traffic signs, road markings and signals

Road markings and traffic signs must be obeyed by horse riders.

Signs giving orders are usually circular. Those that are prohibitive normally have a red edge or are completely red. Signs that are circular and are coloured blue but do not have red borders give positive instructions.

Warning signs are usually triangular with a red border.

Directional signs are usually rectangular. Those on motorways are coloured blue and those on primary roads are green. Signs on local routes usually have a black border.

Horses are not allowed on motorways and it is better, where possible, to avoid main roads.

Road markings. Continuous white lines across a junction or carriageway mean you must stop, whereas a broken single or double line mean you must give way to traffic approaching from your right-hand side.

Signals from authorised persons. You must obey signals from authorised persons such as policemen, traffic wardens and crossing patrols.

Traffic light signals. Horse riders must obey traffic light signals whether they are main traffic lights at a junction or portable lights at roadworks.

The colour sequence of traffic lights is as follows:

- Red means stop.
- Red + amber means stop but get ready to move off; do no go until green appears.
- Green means go if it is clear to do so. If you are turning left or right give way to pedestrians crossing the road. (A green directional arrow means you can go in that direction whatever other colour light is showing *if the way is clear*.)
- Amber means stop. You can only continue if you have already crossed the line when the lights change.

Alternate flashing lights at level crossings, lifting bridges, airfields, fire stations or any other emergency service mean that you MUST stop.

Junctions and roundabouts

- Make sure you know who has priority at different types of junctions.
- Unless you are using a right filter lane, keep to the left at all junctions to avoid being trapped between two vehicles.
- Reassure the horse and do not let him fidget or anticipate moving off.
- Always be aware of the traffic situation.
- Keep looking and listening as you cross the junction.
- Avoid roundabouts whenever possible. If you have to negotiate a roundabout keep to the left. Depending on how busy the roads are, it may be safer to turn left, cross that road, and turn left again onto the roundabout.

Giving hand signals

Horse riders (and those driving horses) need to give hand signals to indicate their intentions to other road users. Make your intentions clear to other road users by giving the correct hand signals and positioning your horse clearly.

Stop well back from a junction so that the horse does not become frightened by passing traffic. Look around to see if you are safe before signalling and making a change of direction.

Signalling routine: observe, signal clearly, observe, manoeuvre. Do not hesitate, be decisive.

Turning left: raise left arm to shoulder level and keep it straight, with a straight wrist and fingers together.

Turning right: raise right arm to shoulder level and keep it straight, with a straight wrist and fingers together.

Requesting cars to slow down: raise right arm and point towards the driver, lift to shoulder level and keep it straight, with a straight wrist and fingers together, raise arm up and down, while looking at the driver.

Request for driver to stop: turn to face driver with right arm outstretched at shoulder height and palm facing driver with fingers pointing to the sky. Hold signal for as long as is needed.

Other points to note

- Any horse on the road must wear a bridle.
- Never signal with the whip in your hand. Hold it with the thumb of your left hand to avoid having to change it over.
- Always expect the unexpected.
- Never move unless it is safe to do so – do not take a chance.
- Give a life-saver look (one last look before moving off).
- Be aware and stay alive.
- Before moving off or turning, look behind to make sure you are safe and then give a clear signal.
- Keep to the left; if leading another horse keep him on your left.
- Follow the flow of the traffic in one-way streets.

Road awareness

Make sure you can control the horse before taking him out on the roads. If he is nervous, ride with a companion.

Try to anticipate what the vehicles around you are going to do.

The faster a vehicle is travelling the longer it will take it to stop once the driver has seen you and your horse. The stopping distances for cars travelling at different speeds are as follows:

20 mph = 12m or 40ft (3 cars' length)
30 mph = 23m or 75ft (6 cars' length)
40 mph = 36m or 120ft (9 cars' length)
50 mph = 53m or 175 ft (13 cars' length)
60 mph = 73m or 240ft (18 cars' length)
70 mph = 96m or 315ft (24 cars' length)

Lorries need more distance to stop. Air brakes on lorries or coaches may go off at any time. This cannot be controlled by the driver so be aware in case your horse becomes upset.

Leading from another horse or from the ground

All led horses must wear a bridle. If leading on foot and the led horse is wearing a martingale leave the reins on his neck. Hold them in your left hand approximately 10cm (4in) from the bit. Carry a whip in your right hand to position the quarters.

Ride on the left-hand side of the road and, if leading another horse, position him between your horse and the kerb so he is shielded from the traffic.

Keep to the left-hand side of the road with the led horse on your left to shield him from the traffic.

THE BRITISH HORSE SOCIETY

(by kind permission of The British Horse Society)

The British Horse Society (BHS) was founded in 1947 and is the nation's leading equine charity with a passion for horses that is backed by knowledge and expertise. The BHS represents and provides a range of services for horse riders, horse owners, enthusiasts and professionals.

The BHS offers the most comprehensive range of equestrian advice, the highest standards of training and qualifications, and campaigns on issues that matter to every horse lover. The BHS ensures that everyone gets the most out of their relationship with horses, and that every horse is protected and cared for.

The BHS helps riders get the most from their riding experience and to feel confident in their riding, horse ownership or business. They offer the highest standards of equestrian training and qualifications; BHS exams are internationally recognised, enabling riders to progress from beginner level to world-class standard in equitation, coaching and stable management skills.

The BHS believes in providing the very best advice, information and support. This can help owners, riders and professionals get the most from their relationship with horses – and improve the welfare of horses and ponies across the country.

When you join the BHS you immediately become part of a wide equestrian community. It's a community that will give you the opportunity to voice your passion. You can become involved by helping as a volunteer, or campaigning on issues that matter to equestrians. These include ending discrimination in the provision of equestrian access and lobbying for more places to ride, preventing abuse and neglect through education, and campaigning for recognition and safer conditions for all equestrians.

A great deal of information is available from the BHS website: www.bhs.org.uk

The BHS Primary Objectives The BHS is a registered charity – numbers 210504 and SC038516 – and its activities are overseen by the Charity Commission. The Patron of the Society is Her Majesty the Queen. The Society's primary objectives, as published in its Memorandum of Association, are:

- To promote and advance the education, training and safety of the public in all matters relating to the horse.
- To promote the use, breeding, well-being, safety, environment, health and management of the horse for the public benefit.
- To promote community participation in healthy recreation involving the horse.
- To promote and facilitate the prevention of cruelty, neglect or harm to horses and to promote the relief, safety, sanctuary, rescue and welfare of horses in need of care, attention and assistance.
- To promote and secure the provision, protection and preservation of rights of way and access for ridden and driven horses over public roads, highways, footpaths, bridleways, carriageways, public paths and other land.

'Horses' for the purpose of the Objectives means any member of the family *Equidae.*

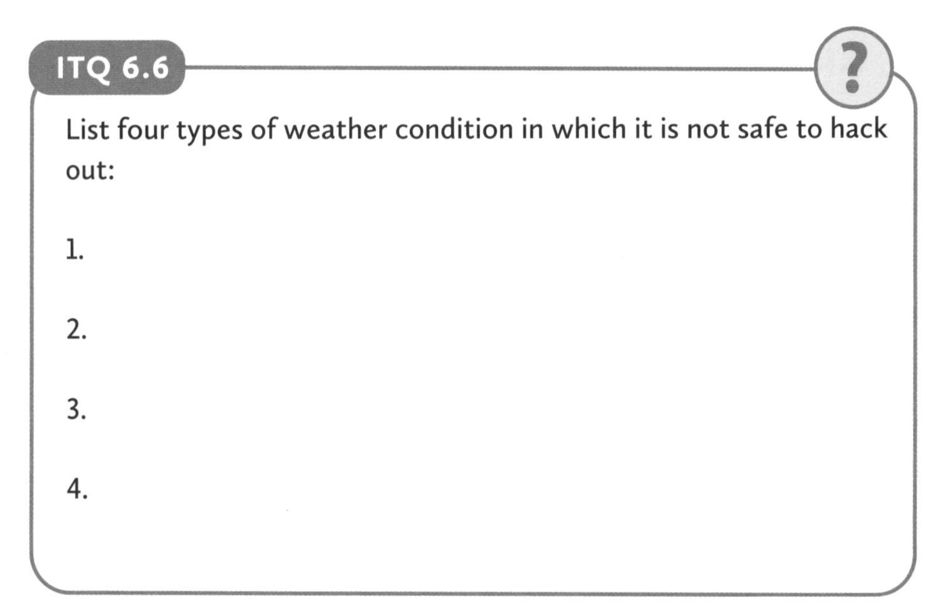

ITQ 6.6

List four types of weather condition in which it is not safe to hack out:

1.

2.

3.

4.

ITQ 6.7

When giving hand signals and turning on the roads, what is meant by a 'life-saver' look?

7 Horse Health and Behaviour

REQUIRED SKILLS/KNOWLEDGE	Learnt, revised, practised?	Confirmed
Learn the signs of good and ill health in a horse and understand basic equine behaviour and welfare.		
• Know what to look for at when making routine morning and evening checks on horses in your care.	☐	☐
• Be able to recognise the signs of good and ill health in a horse.	☐	☐
• Understand the importance of reporting when a horse is unwell.	☐	☐
• Be familiar with natural equine behaviour in the wild and the horse's basic instincts for survival.	☐	☐
• Know the signs of potentially dangerous horse behaviour.	☐	☐

DAILY INSPECTIONS

Constant vigilance is the best way of assuring yourself of horses' well-being, and of picking up early signs that something may be amiss.

The early morning check

You can tell almost instantly if a stabled horse has had a problem in the night by the way he appears first thing in the morning. The horse should appear relaxed, and is likely to be looking out over his stable door in anticipation of being fed. He should then eat his feed willingly.

Indications of problems

- Bedding badly churned up (the horse has been rolling and digging up his bed – possibly a sign of colic).
- Horse in a sweat or showing signs of having sweated (the horse has either been too hot, is in pain, or possibly a sign of colic).

- Rugs completely askew. The horse may have been rolling excessively (possibly as a result of discomfort) which has caused them to slip (or they were not put on correctly).
- Horse has not eaten his food, hay or drunk any water.
- Horse standing at the back of his stable, looking dull and disinterested in his morning feed.
- Horse lying down, not wanting to get up for his breakfast.
- Excessive amount of loose droppings. This may be a result of moving onto new grass, or it may be a sign of a digestive problem.
- No droppings. This is a sign that the horse is suffering from an impaction (this will cause colic).
- Horse resting a foreleg.
- Any new cuts, particularly on the head, could indicate that the horse has been cast in the night.

These basic indicators of potential problems will be expanded upon in Signs of Ill Health, later in this section.

The evening check

When you check the stabled horse last thing in the evening he should have cleared up his evening feed and is likely to be eating his evening hay ration. He may have drunk some water by this time. If it is very late in the evening he may be lying down. Provided he has eaten his feed and appears relaxed, this is normal and he shouldn't be disturbed. However, you need to become familiar with the normal behaviour of each horse.

Signs that all is not right are the same as for the early morning check, with the exception that he may not be looking out over his door as he will know he has had his evening feed and is likely to be eating his hay or resting.

Whatever time of day, if the horse appears unwell you must then assess him for other signs of good or ill health.

ITQ 7.1 ?

How should the stabled horse appear first thing in the morning?

ITQ 7.2

Give two signs that you may notice first thing, indicating a possible problem with the horse:

1.

2.

SIGNS OF GOOD HEALTH

General demeanour and bodily functions

Alert outlook. When fit and well, most horses are alert and show interest in their surroundings. They react to certain stimuli, e.g. feed time, passing horses, loud noises and generally have an interested, yet relaxed attitude.

Appetite. Most horses have a very healthy appetite and clear up their feeds immediately, whereas a smaller number are slow feeders, inclined to be a bit fussy about what they eat. If a usually 'greedy' horse does not finish his feed it could be a sign that something is wrong. 'Slow' feeding can be caused by overfacing the horse with large feeds. (Always remember to feed 'little and often'). The horse may dislike the food or something that has been added to it, e.g. not all horses enjoy cubes or garlic additives.

Skin, coat and mucous membranes. The **skin** should feel supple and quite loose. As you run your hand over the skin you should see small 'ripples' appear. The **coat** should have a smooth and naturally glossy appearance. The **mucous membranes** are the membranes of the gums and eyes. The gums should be moist, slippery to the touch and salmon pink in colour.

Clear eyes and clean nostrils. The eyes must be bright and, like the nostrils, free of discharge, particularly that of a thick, sticky consistency. Occasionally horses have a slight clear, watery discharge – if no other symptoms are present, this can be considered as normal.

Normal excretion/urination. Droppings should be passed regularly and will vary in colour and consistency according to the diet. A stabled horse eating hay and short feed will pass yellow-brown droppings, whilst the droppings of a grass-kept horse will be dark green-brown. The droppings should be fairly firm, breaking apart on hitting the ground and be free of any offensive smell.

Horses urinate approximately four to six times per day, passing between 5–15 litres (approx. 9–26 pints) of urine.

Most horses wait until they are standing over deep bedding or grass before staling as they don't like to do so on concrete or in a lorry without straw or shavings down.

They will often wait until they return to their stable or field.

Horses adopt a typical posture when staling – they stand with their hind legs separated, leaning forwards slightly. Geldings normally extend their forelegs forwards as well and some horses make 'grunting' sounds as they stale.

The urine should be pale yellow to amber in colour and free of any offensive smell.

The limbs. The limbs should feel cool. They should be free from heat, pain and swelling. The horse may stand resting a hind leg, but never a foreleg. He should be happy to stand weight-bearing on all four limbs.

When out of the stable he must be sound, taking even, free steps. The walls of the hooves should feel uniformly cool without excessive cracks and should be well trimmed/shod. The clefts of the frog should be clean and dry without an offensive smell or black discharge.

Temperature, pulse and respiration (TPR)

The normal rates for a healthy adult horse at rest are:

- Temperature — 100.5 °F (38 °C)
- Pulse — 25–42 beats per minute
- Respiration — 8–16 breaths per minute

However, these figures are subject to minor variations between individuals. To gauge what is normal for each individual horse, take the TPR first thing each morning for several days and write down the readings.

To observe respiration

The normal rate of respiration for an adult horse at rest is between 8–16 breaths per minute. The respiratory rate should be taken first as this is unobtrusive and will not upset the horse, which would cause an increased heart rate.

1. The horse must be standing still, at rest.
2. Watch the rise and fall of the flanks. Each complete rise and fall is one breath.
3. Count either the rise or fall for 1 minute.

To take the pulse (heart) rate

1. Press two fingers against the **transverse facial artery**, which is found slightly below and to the rear of the eye, or the **sub-mandibular artery** on the inside edge of the lower jaw, where it passes over the bone fairly near the surface, or:
2. Using a stethoscope – press the stethoscope against the horse's girth just behind the left elbow and listen for the heartbeat. You will hear a 'lubb-dub, lubb-dub…' sound. Each 'lubb-dub' represents one heartbeat.

Count the pulse rate/heart beat for 30 seconds and multiply by two. The pulse rate of an adult horse at rest is between 25–42 beats per minute. Foals will have a pulse rate of 50–100 beats per minute.

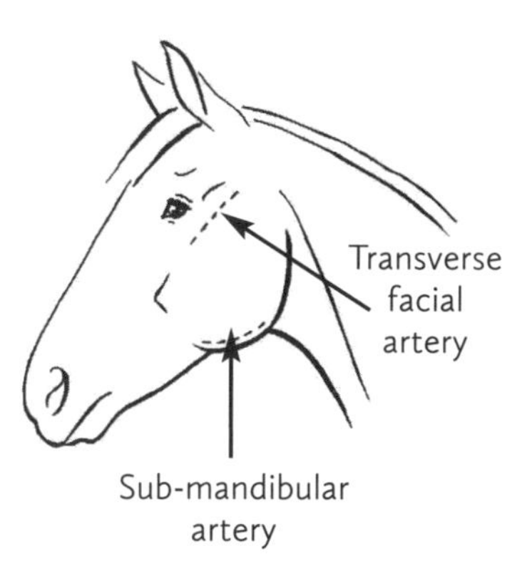

7.1 Sites for taking the pulse

To take the temperature

At Stage 1 level you will not need to take the horse's temperature but it is useful to know how to do it. Have an experienced person on hand to assist when learning to take a horse's temperature.

1. The horse should be untied and held by a competent handler. If tied up he could pull back, break the weak link and jump on your foot should he become upset by the procedure.

2. If using a mercury thermometer*, shake it down so it reads several degrees lower than normal. If using a digital thermometer, switch it on. Lubricate the end with lubricating gel or petroleum jelly. Stand behind the horse, slightly to one side.

3. Hold the horse's tail to one side and insert the bulb of the thermometer. Insert to halfway and hold at a slight angle to press the thermometer against the side of the rectum.

4. Hold in position for 1 minute. A digital thermometer will 'beep' or the digits will stop flashing when the maximum temperature is reached.

5. Withdraw and read the thermometer. Always wipe clean and disinfect before returning to its case. Disinfectant wipes are a convenient way of doing this.

6. Switch off the digital thermometer.

**A digital thermometer is easier to read than a mercury one.*

ITQ 7.3 ?

How should the coat and skin of a healthy horse appear?

ITQ 7.4 ?

Describe how the mucous membranes of the gums should appear.

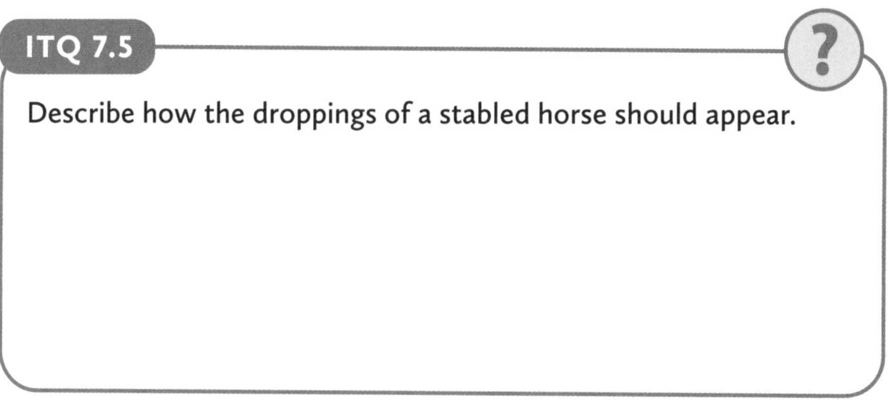

ITQ 7.6

What are the normal resting rates for:

a. Temperature

b. Pulse

c. Respiration

You will develop an eye for health and condition as you spend more time with horses. The most important consideration is to be aware of what is normal for each particular horse. Regularly check the limbs, TPR, etc., so you are familiar with the norm – this way any changes will be more easily recognised.

SIGNS OF ILL HEALTH

Observing and acting

We now go on to discuss the signs of ill health. A horse may show only one or two of the signs listed – it is up to you to be observant and notice the sign(s) in the first instance and then take appropriate action.

If you do notice one or two signs of ill health you should:

- Look for additional signs which will help identify the problem.
- Consult an experienced person who can offer professional help.
- Decide whether or not to call the vet.
- Prevent the condition from worsening whilst waiting for the vet (with the help of your experienced person).

IN-TEXT ACTIVITY

Select a horse known to you and record his temperature, pulse and respiratory rates on consecutive days. Record the details below.

Ask an experienced person on your yard to help you.

	Temperature (°F and/or °C)		Pulse (beats per min)		Respiration (breaths per min)	
Horse's name:	Day 1	Day 2	Day 1	Day 2	Day 1	Day 2

- If you decide the vet is not needed:
  — identify the cause
  — treat the problem
  — take steps to prevent it from occurring again.

Common signs

Disinterested, dull attitude

Standing in the corner of the stable with the head low is one of the first signs to indicate all is not right.

Possible causes

- The horse *could* simply be tired and having a rest! Keep a close eye on him though.
- A viral or bacterial infection could be starting.

Loss of appetite

If the horse fails to clear up a feed and is not interested in his hay or water, you should be on the lookout for further symptoms.

Possible causes

- He may not like an additive, e.g. garlic, or a wormer that has been added to the feed.

- A horse new to a yard may take a day or two to settle in and start eating properly.
- A viral or bacterial infection could be starting.
- The horse may be starting to have abdominal pain (colic).

Dull and staring coat

Note that unrugged, grass-kept horses can look dull when changing from their summer coat to their winter coat. This is perfectly normal and provided the horse appears well otherwise, is not a sign of ill health. In other circumstances, however, a coat like this may indicate that all is not well.

Possible causes

- The horse may be cold.
- Worm infestation.
- Nutritional deficiency.
- General lack of condition.

Tight skin

Dehydration causes the skin to lose its elasticity and feel 'tight'. Dehydration is a serious condition and, if suspected, needs to be investigated thoroughly.

Sweating

This needs to be investigated in context, for example the horse may sweat if he's just been exercised (particularly if not clipped appropriately), or he may be wearing too many rugs/too heavy a rug in mild weather (many people tend to over-rug). If these factors are ruled out, other possibilities need to be investigated.

Possible causes

- The stable (or lorry/trailer) may be stuffy and poorly ventilated.
- Excitement.
- Pain.

Having ruled out obvious reasons for sweating, look out for other signs, especially **colic signs**. These include:

- Pawing at the ground.
- Looking round at the flanks/kicking at the belly.
- Repeatedly lying down and getting up.
- Rolling.
- Lying flat out in the stable.

- Increased pulse and respiratory rates.
- Loss of appetite.

SAFETY TIP

- If a horse is rolling and thrashing out in pain there is a high risk of serious injury to anyone who enters the stable – an experienced handler only should enter the stable with extreme care, if at all.

ITQ 7.7

Give two reasons why a horse may not eat his feed:

1.

2.

ITQ 7.8

If a horse in your care looks dull, compared to normal, what action would you take?

ITQ 7.9

Why should you keep a record of the horse's normal TPR rates?

ITQ 7.10

Give two signs that may indicate a viral or bacterial infection is starting:

1.

2.

Lack/loss of condition

This is fairly easy to recognise. A horse in poor condition usually has a dull coat, does not carry enough flesh and may even have projecting hips, shoulders and ribs if he is very thin.

Possible causes

Horses don't normally lose condition overnight so poor condition indicates an ongoing problem which needs to be dealt with. The more common causes include:

- Poor diet – insufficient food.
- Sharp teeth which prevent the horse from chewing his food properly.
- Worm infestation.
- Overwork.
- Cold, e.g. thin-skinned horse kept out with inadequate shelter.

A rapid loss of condition indicates a serious problem and the vet should be consulted.

Abnormal discharge from eyes and/or nostrils

Possible causes

- Discharge from the eyes and nostrils indicates an infection such as a cold or influenza, especially if accompanied by a 'wet' sounding cough.
- Discharge from the eyes only, especially if the lids and/or membranes are inflamed, indicates a foreign body such as a hay seed, or a condition such as conjunctivitis.
- If the horse is allergic to the dust found in hay and straw he may have a thick yellow (mucous) nasal discharge accompanied by a dry cough.

Abnormal mucous membranes

Possible causes

If the membranes of the eye and gums are:

- Pale coloured – anaemia.
- Yellow (jaundiced) – liver complaint.
- Tacky and dry – dehydration.

Abnormal droppings

Possible causes

- Too hard – constipation.
- Too soft/loose – worm infestation, excitement, too much rich grass, sharp teeth preventing the horse from chewing his food properly.

- Diarrhoea – infection such as salmonella, or poisoning.
- No droppings – constipation or impaction (blockage).

Abnormal urine/urination

Possible causes

- Thick and cloudy and/or bloodstained – kidney disease.
- Smell of violets and dark-coloured – azoturia.
- Repeated efforts to urinate without producing any urine – kidney problems, cystitis.

Abnormal limbs

Possible causes

- Cold swellings are often a consequence of poor circulation caused by lack of exercise. Once the horse has been turned out or exercised this type of swelling often goes down.
- Hot swellings may be a result of impaired circulation, infection or injury, e.g. a sprain. The swelling may or may not be painful, and the horse may be lame. The vet should be consulted.
- It is normal to rest a hind leg, but never a foreleg. A pottery, stilted action indicates laminitis. The laminitic horse will also stand with the weight back on his heels and be reluctant to move. He may shift his weight from one forefoot to the other.

Heat in one or more hooves may indicate a problem such as infection or laminitis.

Abnormalities of TPR

Possible causes of abnormal temperature

- A rise of one or two degrees indicates pain, e.g. colic or injury.
- A rise of more than two degrees indicates a more serious infection.
- Hypothermia will cause a reduction in body heat.

Possible causes of abnormal pulse rate

- A rapid pulse of 43–50 beats per minute indicates pain.
- A very weak pulse indicates that the heart is failing, e.g. when a horse is in shock.

Possible causes of increased respiratory rate

- Pain.
- Laboured breathing and respiratory distress indicates damage to the lungs, e.g. chronic obstructive pulmonary disease.

When investigating TPR abnormalities, take into account that a horse who has just worked strenuously will have increased temperature, pulse and breathing rates.

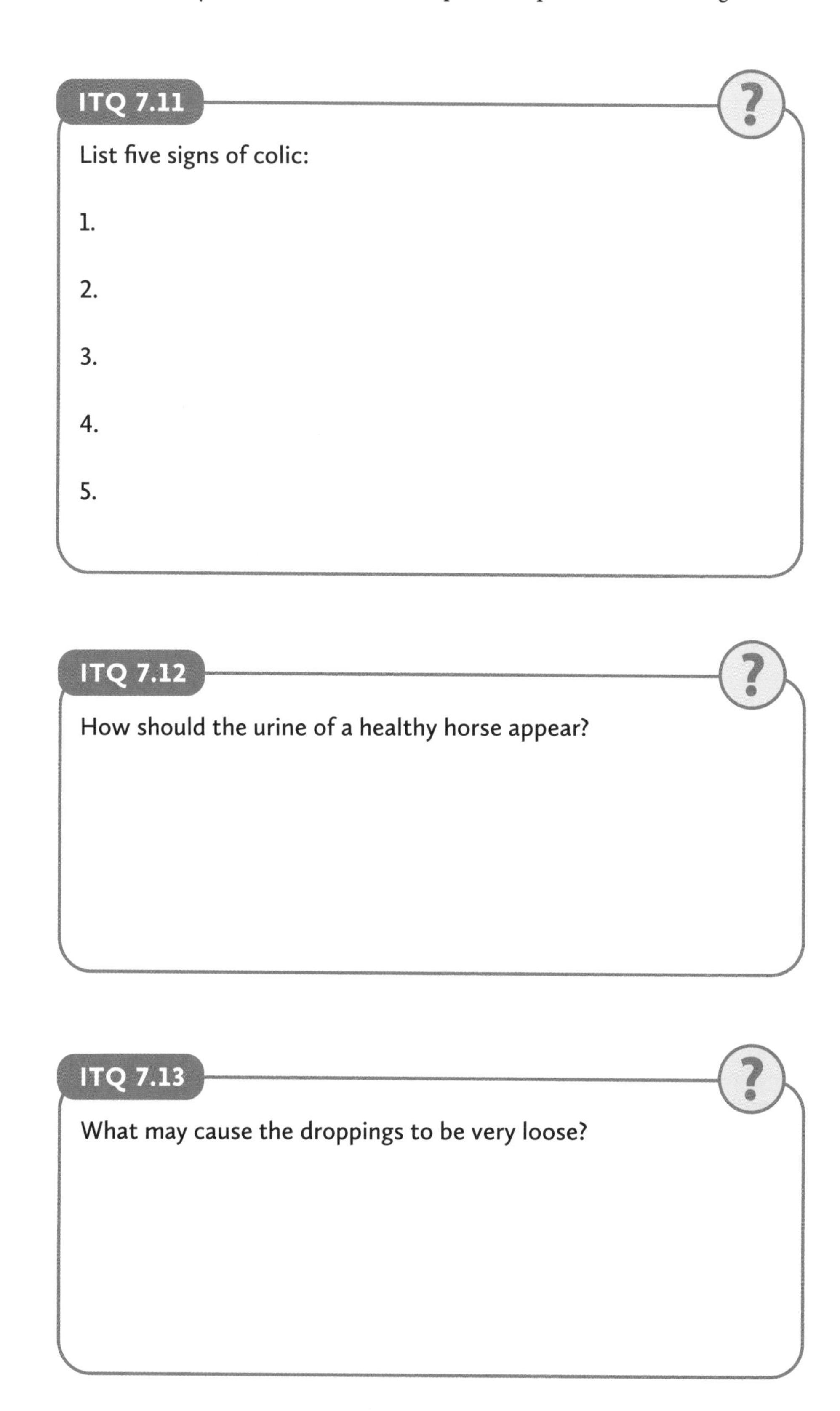

ITQ 7.11

List five signs of colic:

1.

2.

3.

4.

5.

ITQ 7.12

How should the urine of a healthy horse appear?

ITQ 7.13

What may cause the droppings to be very loose?

ITQ 7.14

Give four causes of poor condition:

1.

2.

3.

4.

ITQ 7.15

What ailment causes the horse to move with a pottery, stilted action?

Wounds

Any break in the skin, i.e. a wound, needs to be assessed and treated. Simple wounds can be cleaned and, if necessary, protected. Check that the horse is vaccinated against tetanus. More complicated wounds may need to be stitched or stapled, requiring veterinary attention.

The heel region should be checked for cracking and mud fever in wet conditions, particularly if the horse is turned out in a muddy paddock.

When to call the vet

Report any abnormal observations to the yard manager immediately; delays can cause a condition to worsen. If no one else is present you will need to make a decision whether to call the vet or not. Phoning the vet and discussing it will also help clarify whether the vet is needed.

Keep the vet's telephone number displayed near every phone to minimise delay. Call the vet out for any of the following situations.

- **Suspected colic.** Call the vet out if the horse shows mild signs of colic for 20 minutes or more, or immediately if the horse shows violent signs. The vet will give pain-killing and muscle relaxant injections as even a mild case of colic can soon worsen; a horse in great pain will roll and thrash about, risking further injury and life-threatening complications such as a twisted gut, which requires surgery.

- **Wounds.** Call out the vet if a wound is:
  - Very deep
  - Complicated, e.g. on a joint
  - Infected
  - In need of stitching
  - Bleeding profusely
  - Spurting blood (indicating arterial bleeding)
  - Has punctured the sole of the foot.

The vet will also be needed if there is any doubt about the horse's tetanus vaccinations. A quick-acting tetanus anti-toxin will be needed.

- **Lameness.** If a horse is lame and you cannot determine the reason.

- **Suspected fractures.** Accidents can happen whilst horses are turned out to grass or being ridden.

- **Repeated coughing.** The horse may or may not have a purulent discharge from one or both nostrils.

- **Abnormal temperature.** A variation from normal of more than 1 °F.

- **Suspected laminitis.** This is such a serious and painful condition; the vet must be called as soon as a horse shows signs.

Other than this, a general guide is to call the vet when your horse is showing any of the signs of ill health and you are unable to determine why, or to administer the necessary treatments.

The horse should not be worked if showing signs of ill health.

Preparing for a veterinary visit

Whenever the vet is to visit your yard try to ensure that:

- If possible, the horse is in a dry, well-lit loose box.

- There is warm water, soap and towel available for hand-washing before and after treatment.

- If it could be relevant, any droppings the horse has passed should be kept in a skip. This may help the vet to make a diagnosis.

- A record is kept of all relevant information, including any food the horse has eaten, and when, in the previous few days, as well as any other signs of abnormalities.

If in any doubt, it is always better to play safe and call the vet out. A vet would far rather come out and deal with a minor ailment, than have to attend to a problem that has been left and worsened, possibly causing the horse to suffer in the meantime.

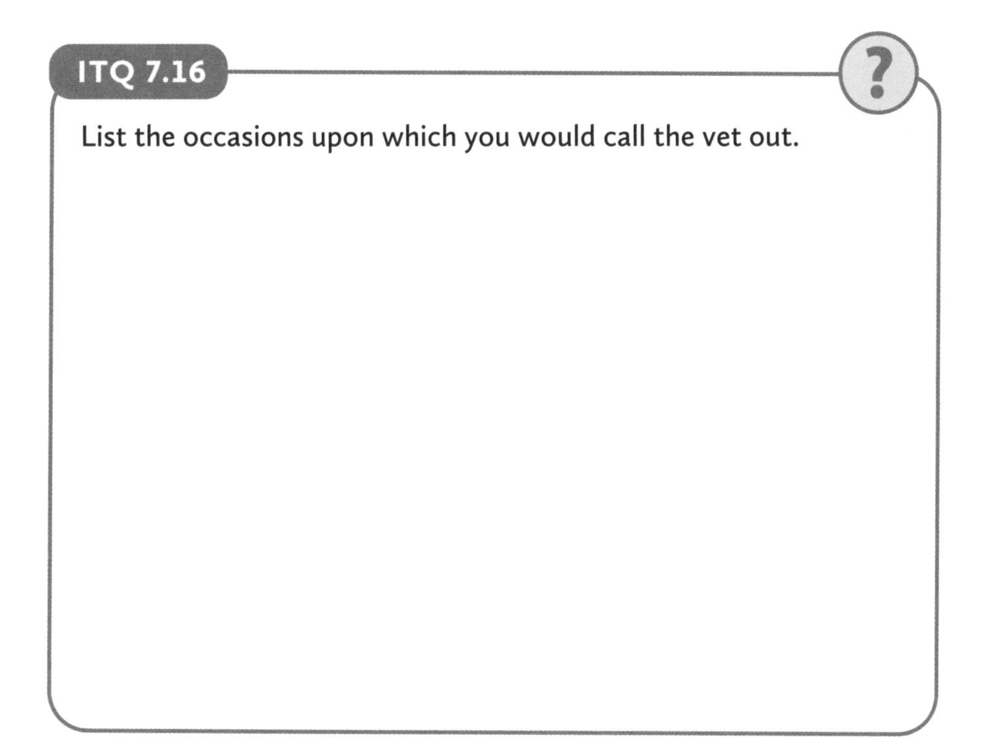

HORSE BEHAVIOUR

Instinctive behaviour

When handling horses regularly you will notice the various ways in which they react to different situations. Most reactions are instinctive and an understanding of why horses react in certain ways is gained through a basic knowledge of the main instincts of the horse.

Instinct is an inborn and natural automatic reaction to stimuli. These instincts are so strong that they survive to this day within our domesticated horses.

The sensory systems

Our domesticated horses' ancestors were wild creatures of the plains who roamed in herds. Their main instincts were essentially geared towards survival. Survival of a prey animal depends on an efficient warning system, provided by the sensory systems.

Sight

Because of the lateral position of the eyes, the horse has a wide visual range, even when grazing. This allows the horse to see predators approaching from behind, as well as being able to see the rest of the herd without having to lift his head. The horse's eyes are very large to allow a large surface area of light receptor cells. This, combined with an abundance of specialised rod cells, gives the horse very good night vision.

As the horse's eye also contains cone cells (those responsible for identifying different colours), it is thought that the horse can differentiate between colours.

Hearing

The external ears are mobile enough to be able to catch distant sounds and direct them into the auditory canals. Horses can discriminate between different words and are able to react to the inflection and tone of the voice.

Smell

The nostrils are able to detect edible feedstuffs, fresh water and airborne scents. Pheromones are 'smell messages' produced by the glands of the body and can be detected in the breath, sweat, urine and faeces. Stallions can smell in-season mares, whilst all horses use their sense of smell when 'getting to know' each other. Mares recognise their own foals through smell and vice versa.

Territory can be 'marked' and recognised through smell. Stallions and some geldings will always urinate and defecate in one area of their field. Groups of horses can take on a 'communal' pheromone through close contact, e.g. through mutual grooming and rolling in the same place.

Taste

This sense is closely associated with smell and helps the horse to avoid eating harmful and unpleasant plants – although, through desperation, a starving horse may eat poisonous herbage. However, we must not rely on the horse to discriminate between harmful and non-harmful feedstuffs – many poisonous plants do not smell or taste unpleasant and the horse may inadvertently ingest them with dire results. Some horses actually relish acorns, which are poisonous.

Feel

The whiskers act as feelers which help to prevent unwanted objects from being eaten. The skin is sensitive to touch so an unwanted presence can be dislodged – the natural reaction is to remove a predator and flee, usually by bucking or scraping under a tree or similar. It is this same sensitivity that allows a rider to train a horse to respond to very subtle aids.

Herd membership

When considering how horses live in the wild you must take into account that they are naturally gregarious creatures who instinctively feel 'safety in numbers', resulting in a strong desire to stay within the herd.

Horses naturally group together as a way of increasing their chances of survival – the more there are in the group, the greater the chance of spotting and evading predators. In the event of attack, the individual's chances of survival are increased according to the size of the group.

When food becomes scarce, groups may split up in order to seek out food in different areas. In the wild, horses mainly use their sense of smell to seek fresh water and herbage, eating small amounts as they roam, i.e. they are browsing herbivores. Over time, the horse's digestive system has developed in such a way that only small quantities can be contained in the stomach, while the hind gut (large intestine) has the capacity to hold large amounts of roughage which may take several days to break down and be fully digested.

This must be taken into consideration when feeding the domesticated horse. It must also be remembered that horses have this instinct to roam and graze – the stabled horse will benefit greatly from periods at grass each day, giving him the opportunity to act instinctively.

Care must be taken when feeding a group of horses at grass that a timid animal is not bullied. He will surrender his food to a more dominant, aggressive horse.

Horses kept in groups will interact with each other and form friendships within the group. Certain horses will get on well, whilst others will show aggressive behaviour towards each other. The relationships between horses are complex – just as with humans, different personalities affect the way in which individual horses relate to each other.

It can be stressful for a horse to be separated from the other members of his herd or group, as he may feel threatened; for example a horse left alone in his stable or paddock when all the others go out on exercise may become very upset. He may pace frantically, whinnying loudly and sweating up. In extreme cases he may try to jump out.

A nappy young horse is acting upon the instinctive feelings of danger when leaving the herd. Youngsters will often 'nap' (refuse to go forwards, possibly rearing, whipping round suddenly or simply standing rooted to the spot). This has to be dealt with firmly and fairly, rewarding the horse when he has done as asked.

The trainer must instil confidence, respect and trust in the horse. The horse must be taught that it is safe to leave the herd; that dangerous behaviour will not be tolerated and that good behaviour will be rewarded. This does not, of course, happen overnight!

The herd instinct plays a role in the stable yard – the stable often becomes the horse's 'safe house', a substitute for the herd. This may explain why horses tend to walk out more freely and are less likely to nap on the way home from a hack, as they are keen to get home. A frightened horse instinctively returns to his stable or herd for safety. This may be seen when a horse breaks loose or when a rider has fallen off. The horse usually gallops back to the yard or, if at an event, to his lorry, and tries to go to his companions; his fellow 'herd' members.

It is this sense of security within the stable than can be so dangerous in the event of fire – the horse smells and hears burning and is afraid. The stable is his security, his safe place, and he will not want to leave. The horse will often need to be disorientated by covering his eyes and nose with a damp jacket or towel, before he will allow himself to be led to safety.

At grass, horses in a herd will enjoy mutual grooming and in hot weather often stand nose to tail swishing the flies from each others' faces. It is normal for horses in a field to basically stay together. It is not normal for one horse to become totally separated from the others for any length of time.

One horse will often remain standing 'on guard' whilst the others lie down and sleep.

Horses will often play, cantering around together. Geldings often play-fight as they would in the herd as young colts – some geldings are more aggressive than others and this play-fighting can become dangerous, risking injury.

One of the older mares in a herd will become the alpha mare, i.e. the herd leader, and can often be seen 'rounding up' the herd, especially those she considers younger. The mare will often stay between the herd and any new horses or ponies that are subsequently turned out in the same field.

As herd membership is so important to horses, it is not ideal to keep a solitary horse as they do not enjoy living alone. If no other horses are available, a goat or sheep should be provided for companionship.

The defence system

Once a horse has been alerted to danger the hormones **adrenalin** and **noradrenalin** are released in response. These prepare the horse for flight or fight, hence they are known as the **'fright, flight, fight hormones'**.

- **Fright.** Having ascertained that danger is imminent, the horse utilises his defence system.
- **Flight.** The horse has evolved as a fast animal whose long legs allow him to gallop away from danger. Horses tend to gallop back to the herd to escape predators.
- **Fight.** The horse can defend himself by:
  - Striking out with either fore or hind feet.
  - Bucking and rearing to remove the predator.
  - Squashing and scraping the predator off under a branch.
  - Biting.

Horses are naturally wary of snakes and sudden, sharp movements. Many horses, particularly youngsters, are initially frightened by a hosepipe being dragged carelessly beneath them or of a trailing lead-rope or lunge line. Most horses will shy away from sharp, sudden movements as a means of protection.

ITQ 7.17 ?

Give four examples of ways in which horses demonstrate the herd instinct:

1.

2.

3.

4.

ITQ 7.18

a. Which hormones are known as the 'fright, flight, fight hormones'?

b. How can a horse defend himself if he has to fight?

Demeanour and communication

Horses use a wide range of signals to communicate how they feel both to fellow horses and their human handlers. The whole body is used but it is the facial expressions in particular which indicate how a horse is feeling.

A relaxed horse will seem calm and generally at ease with his companions and surroundings. If resting either in the field or stable his head may be lowered and in some cases, the lower lip may hang. His eyes may be half-closed; he may seem drowsy and may rest a hind leg.

An interested horse will appear bright-eyed with an inquisitive, interested expression, ears pricked forward. He will raise his head and appear alert.

An excited horse is likely to carry his tail high, prance about whether ridden or in the stable, and his ears will prick to and fro. Eyes and nostrils will be wide open to take in the sights and smells that are exciting him. He may 'snort' and vocalise (whinny). Often horses will sweat up and defecate or stale in their excitement. These signs are often seen in young horses at their first show or out hunting for the first time or in a group of horses when the hunt goes past their yard or field.

A frightened/nervous horse may break out in a sweat and appear physically very tense, with his head raised and back hollow. He will move his ears in an attempt to detect the sounds of danger and flare his nostrils and 'snort' to pick up smells.

His eyes could be described as being 'out on stalks' as he becomes wide-eyed to try to see the danger; he may 'roll' his eyes to gain a better lateral rear view. This expression indicates that the 'fright, flight, fight' hormones have probably been released so the horse may resort to one or more of his defence mechanisms. You may hear or feel (if riding) his heart literally pounding, much faster and stronger than usual.

The most common reaction of a frightened horse is to take flight – he will be unwilling to approach the object of his fear and may try to move away in the opposite direction, normally at speed. Although it is the survival instinct kicking in, at such times most frightened horses do not have a sense of what is going on around them and no perception of the danger their flight response will engender.

As a handler or rider you must be aware that a frightened horse will behave very erratically, and can be a danger to himself and you.

When expressing **dominance** or **aggression** the horse may show the whites of his eyes, lay his ears flat back and bare his teeth, giving an aggressive, dominant expression. This can often be seen when a group of horses are fed in a field – there is a pecking order and the timid horses surrender their feed to the dominant ones: a factor which must be considered when feeding grass-kept horses. The horse then often turns his hindquarters to the others, threatening to kick.

When horses show aggression they may vocalise – this can sound like a deep-pitched 'roar' or high-pitched squeal.

Communication and the ridden horse

Horses also use their 'body language' when being ridden. Tension and discomfort may be expressed through ears laid back, grinding teeth, tail-clamping or swishing, arching of the back, unsteadiness in the mouth, bolting, rearing and napping. Horses may buck either through high spirits or in a deliberate attempt to deposit the rider – it is usually fairly obvious which it is!

The rider must determine why a horse is showing signs of tension or behaving badly:

- Is he in pain? For example from back or tooth problems or ill-fitting tack?
- Does he lack confidence? Is he being asked to do more than he is able, both physically and mentally?
- Is the handler consistently firm and fair? Does the handler have the expertise to train this horse?
- Is the horse taking advantage of the trainer's lack of expertise and 'getting away with it'?

These are some of the questions to be asked under such circumstances. Having found the answers to these questions, steps can be taken to rectify the problem.

HANDLING HORSES

By observing horses and their reactions to various stimuli, you can gain a greater understanding of what makes them 'tick'. An understanding of the instinctive reactions of the horse helps us to appreciate why horses behave in particular ways. This understanding should be used to build up a sympathetic and trusting rapport with every horse you handle and train.

It is important to handle horses in a quiet and confident manner, making sure you are always aware of how the horse is reacting. You need to be aware of problematical situations that may arise when handling horses in the stable, yard or field.

Dealing with problems

Biting and snapping

Although horses sometimes bite without warning there are some signs to look out for. These vary from laying the ears back, tail-swishing and threatening to bite to the more dangerous habit of lunging at you from the back of the box with teeth bared and making every effort to bite and are often caused, originally, by mismanagement particularly in the early stages when youngsters nip playfully and are not reprimanded.

Some horses are very sensitive and thin-skinned, or even just ticklish and dislike being groomed or rugged up. Feeding titbits, particularly over the stable door, can encourage horses to bite.

Control

- Treat the horse fairly and firmly, reprimanding him quickly if he actually bites you. A quick slap on the side of the muzzle is normally enough punishment. Beware of making the horse head-shy through repeated punishment. The horse may also pull back violently when you reprimand him.
- Grooming and rugging up must always be carried out carefully. When tightening girths and rollers take care not to pinch the skin.
- If you feed titbits, do so only after a period of work, or put them in the feed bowl at feed time.
- A confirmed biter must be tied up when being groomed or rugged up, and should either be muzzled, have a grille on the door or a notice on the stable warning people of the danger.

Kicking the handler

This is a very dangerous vice because of the damage a shod foot can do. Like horses who bite, kickers must be dealt with firmly and effectively. Turning the hindquarters to you, not allowing you to approach his head, is a potential kick warning, especially if the horse's ears are laid back.

The causes are also very similar to biting; some horses kick out when being groomed or rugged up because they are sensitive or very ticklish. Some may do it because they have been badly treated and are using it as a form of defence because they are nervous about what they are going to be asked to do.

Control

- Tie the horse up when grooming, tacking up or rugging up.
- Stay close to his body at all times to minimise the chance of a shod foot striking you.
- When grooming around the quarters and hind legs have an assistant hold up a foreleg, and grasp the tail yourself, holding it to one side.
- Be very considerate when grooming and use a soft brush reasonably firmly so you do not tickle the horse.

- Be positive in the way you handle the horse generally and do not put yourself in a position where the horse can make contact with you with any of his four legs.
- When exercising a horse who is known to kick, tie a red ribbon in the top of his tail as a warning to others.

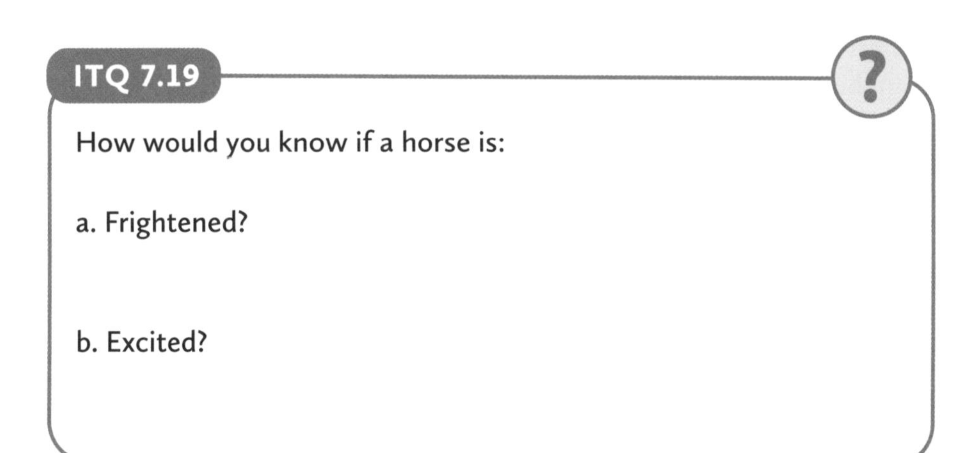

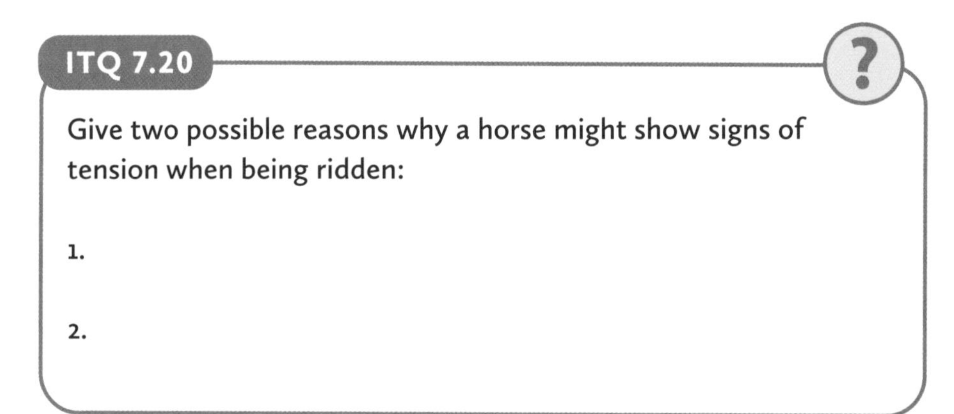

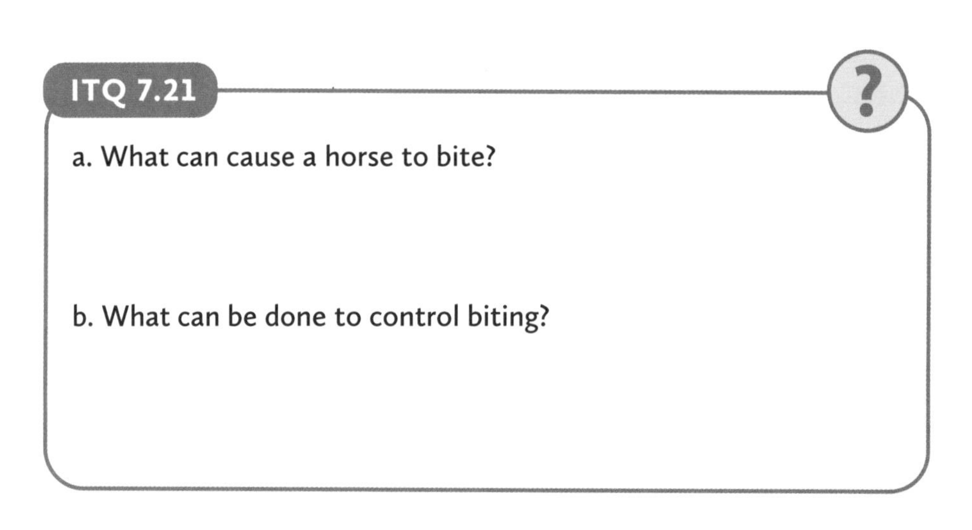

Rearing

Rearing is generally considered to be the most dangerous of the vices – the horse may rear when ridden or being led. Horses rear for one or more of the following reasons:

Disobedience – it often starts when the horse shows signs of nappiness – an unwillingness to go forwards which is not corrected immediately. The horse may start with half-rears and get progressively higher as the nappiness goes uncorrected.

Pain and/or **fear** – rearing is still not acceptable, even if it is occurring as the horse's way of expressing pain or fear, but the reason for the rearing must be identified and dealt with.

High spirits – youngsters, particularly stallions, have a tendency to be high-spirited, especially at their first show, for example when being led during in-hand showing classes.

Confinement/restraint – a horse who is confined or restrained, for example, in order for a treatment to be carried out, may rear because he cannot move forwards.

Control

- Have the horse's back and teeth checked to ensure that he is not suffering from pain or discomfort. Check the fitting of the saddle and bit. An over-strong/severe bit may cause a horse to rear as he tries to evade it.

- Prevention in the first place is better than control. Horses must be trained to go forwards when asked and any nappiness must be immediately corrected with firm and positive riding.

- As a Stage 1 level rider you should never be in the position where you are on a horse known to rear. Horses known to rear should only be ridden by an experienced person – someone who is able to deal with the problem. If a horse does rear the rider's bodyweight must be brought well forward to help prevent the horse from going over backwards. As soon as the horse is down on all fours again he should be reprimanded with the leg and whip immediately and with one hundred per cent conviction. The horse must be made to go forwards. Should he attempt to rear again use one rein to try to turn him to one side – this makes it difficult for the horse to go up – and persist in sending him forwards.

- Most nappy horses will take advantage of an inexperienced or nervous rider. Rearing when ridden is such a dangerous vice that it is negligent to allow a novice person to ride a known rearer.

- When leading, a Chifney or stallion chain gives more control than a normal bit and a longer lead-rein should be used.

- There are other techniques used by different trainers to cure rearing which are beyond the scope of this book and are best left to those who specialise in re-training problem horses. The main principle of the rearing horse is to keep him going forwards and reprimand immediately he shows any inclination to rear. As mentioned, as a Stage 1 candidate you should not be riding horses with such problems!

Bucking

Horses buck either as a way of expressing high spirits or in a determined effort to remove the rider. (Bucking can also be a pain response – as mentioned later under Control.) The high-spirited buck is normally easy to sit to and can be controlled by keeping the horse's head up and riding him forwards positively.

The horse who bucks seriously will normally take off suddenly, head down, twisting as he bucks. 'Fly bucking' is the term used when the horse leaps forwards off the ground, bucking as he goes. Signs that the horse may be about to buck range from feeling the horse's 'back is up', i.e. his back is tense, not relaxed, to a positive arching of the back and lowering of the head. If you feel this happening beneath you, send the horse forwards positively and do not allow him to lower his head.

Control

- Find out why the horse bucks – it may be a result of pain or discomfort. Have the horse's teeth and back checked to rule out discomfort. Check that the saddle fits correctly.
- He may buck because of lack of training – the young horse often bucks until he learns to accept the saddle and rider. Young horses tend to be livelier and more high-spirited than mature ones. Schooling and maturity will improve this.
- Horses who are overfed and underworked will have excess energy to burn. They normally do this by misbehaving, bucking and shying when ridden. Reduce the energy feeds and increase the workload to solve this type of bucking.
- The 'problem' bucker – there are horses who buck for no good reason at any particular opportunity. Check out all the other possible causes first and enlist the help of an experienced rider. To make it more difficult for the horse to buck, keep him working on a contact – do not allow him to have a loose rein. Working him on a contact allows you to keep control of his head – horses normally (but not always) need to get their head down to buck. As with rearing, inducing lateral bend also makes it harder for the horse to buck. Generally, ride the horse forwards positively and make him work hard to take his mind off messing about.

ITQ 7.22 ?

Give three causes of bucking, and their remedies:

1.

2.

3.

8 Grazing

REQUIRED SKILLS/KNOWLEDGE	Learnt, revised, practised?	Confirmed
Maintenance of a safe grazing environment.		
• Know what to look for when checking a field every day.	☐	☐
• Acceptable safe methods for turning out, handling and catching a horse at grass.	☐	☐
• How to recognise, avoid and improve a horse-sick field.	☐	☐

SAFETY AT GRASS

- Never use barbed wire as fencing for horses at grass. Very serious injuries are caused when horses become entangled in barbed wire.
- Check fencing materials for safety – broken rails need to be promptly repaired and protruding nails removed. Check that no rails protrude into the field, e.g. slip rails must be fully opened before leading horses through. Slip rails must never be left partially closed when horses are loose in adjoining fields. Serious injuries can result if the horses canter past protruding rails.
- The fence should be high enough to deter the horses from jumping out. The top rail should be at least 1.2m (4ft) high. The bottom rail should be high enough to stop a horse putting a foot over it but not so high that a pony or foal could roll underneath it.
- If plain wire fencing is used, a wooden top rail must be used to help the horses see the fence when galloping around. It also deters them from leaning on the fence and causing the wire to sag.
- Serious injuries can be caused by horses getting a foot caught in the gate. Ideally gates should have vertical bars or a wire mesh base. Metal five-bar gates are particularly dangerous as the horse can get a foot trapped in the 'V' shaped sections between the bars.
- Horses are particularly prone to injuries when they are in adjoining fields as they may fight at the fence. This is when they are more likely to put a foot through the fence or gate, or try to kick each other. Use a strip of electric fencing along the top of the rails to keep the horses away from the fence. Double fencing of fields with

a gap between is costly, but is another way of keep horses in adjacent fields a safe distance apart.

- Never leave a horse out in the field wearing a nylon headcollar. If the headcollar became caught up, it won't break and could cause serious injury.
- Make sure the water trough doesn't have any sharp protrusions.
- Trim back sharp branches on hedges and trees.
- Fill in rabbit holes as a horse can be injured if he puts a foot down one whilst cantering in the field.
- Turn horses out in small groups known to get on with each other. The smaller the group, the less likely they are to fight.
- Introduce new members to the group carefully. Let the horses meet in the yard before turning them out together. Keep a close watch on them and remove one or the other if they don't get on. In particular you should check for bullying.
- When turning out a horse for the first time, e.g. after box rest or similar, use the smallest paddock to prevent him from galloping around too energetically. Put brushing and overreach boots on if the horse is likely to charge around at first.
- When turning horses out, once the gate is closed lead them a short distance into the field, and, keeping the horses apart, turn them to face the gate. Everyone should undo headcollars and release the horses at the same time, stepping back as they do so to avoid being kicked.
- Always keep fields tidy. If show jumps are kept in the field they should be neatly stacked away for safety when not in use. They should not be allowed to lie around the field as a horse may injure himself.

THE FIELD

Size

As an approximate guide, **0.4 hectares (one acre) per horse** is the minimum land requirement. There are several factors which affect the exact amount of land needed. These factors include:

- The quality of the grass. If the grass is very good and well looked after it may be possible to manage on less land.
- Whether the horses are to live out permanently or to be stabled for some of the time. If the horses are to be partially stabled it is possible to manage on slightly less land per horse.
- Are there any ponds or boggy patches in the field? These will need to be fenced off for safety, which will reduce the amount of land available to your horses. Well-

drained land is more productive and can be used all year round.

- Are you going to ride on the fields as well? Riding on fields cuts up the ground and kills the grass – this will reduce the amount of grass available for the horses to eat.
- The field should be divided up so that sections can be rested to allow the grass to recover and grow.

Fencing

All fencing used with horses must be solid, safe and high enough to deter them from jumping out. i.e. a minimum of 1.2m (4ft) high.

Post and rails

If well maintained, sturdy hardwood post and rails fencing is the safest form for horses. It is expensive but will last for a long time, so is a good investment.

The posts and rails must be treated with preservative to stop the horses from chewing the wood and to prevent the fencing from rotting. Three rails are ideal but two will suffice. The rails should be nailed from the inside of the field to prevent the horses from loosening and dislodging the rails if they lean on the fence. Check the fencing regularly for projecting nails and broken rails.

The lower rail must be high enough from the ground to prevent a horse putting a foot over it, but not so high that a small pony or foal could get underneath it.

Plain wire fencing

Whilst not ideal, two strands of **plain wire** can be used provided a wooden top rail is also used. The wooden top rail allows the horses to see the fencing from a distance, which will prevent them galloping into it by mistake, and also prevents the horses from stretching the wire by leaning over the fence. The main problem with plain wire is the risk of injury should the horse get a foot over the wire.

Barbed wire *is very dangerous and totally unsuitable as fencing for horses – it should never be used.* It causes serious injuries when horses get a foot through the wire or gallop into it.

Electric fencing

Electric strip fencing is economical to buy and is suitable for dividing fields or to act as an extra fence to separate horses in adjoining fields. It is important that the battery is kept well charged or that the power comes from the main electricity supply – as soon as the horses realise the fence is no longer electrified they tend to walk through it.

It can also be used along the top of post and rail fencing if chewing is a real problem.

Hedging

Natural hedges make good fencing provided they are very thick. Gaps should be fenced with post and rails – *not* barbed wire, as is often seen. The ultimate fencing for horses would be hedging lined entirely with post and rails – this provides safe, secure fencing and the hedging provides a good wind-break. Non-poisonous hedges must be used, e.g. hawthorn and beech. Yew, privet and box must *not* be used as they are all poisonous to horses.

ITQ 8.1

Give two examples of accidents that can happen whilst the horse is turned out in the field:

1.

2.

ITQ 8.2

List three safety measures that can be taken to prevent accidents in the field:

1.

2.

3.

Shelter

Shelter must be provided as protection from:

- Winds, especially (in the UK) of north-easterly origin.
- Flies and midges.
- The heat of the sun.
- Severe winter snows.

You need to check with the local planning authority to find out if planning permission is needed before erecting a shelter.

A large open-fronted three-sided shelter is ideal. It should face south-west (in the UK) and have a well-drained non-slip base, which should extend enough to prevent excessive 'poaching' (churning up of ground) around the entrance. The opening must be wide enough to stop a timid horse being trapped in the shelter by a more dominant horse. Alternatively, a straight screen can be built into the fence line, or a double-sided or multi-angle screen used to provide a wind-break. If it is not possible to erect a shelter, good, dense hedging or a clump of trees will provide some shelter.

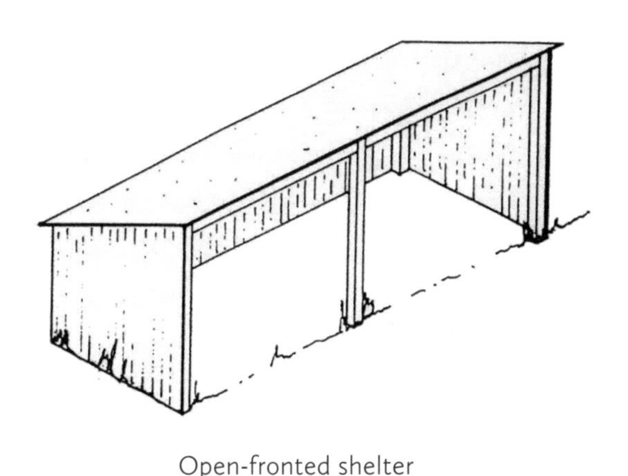

Open-fronted shelter

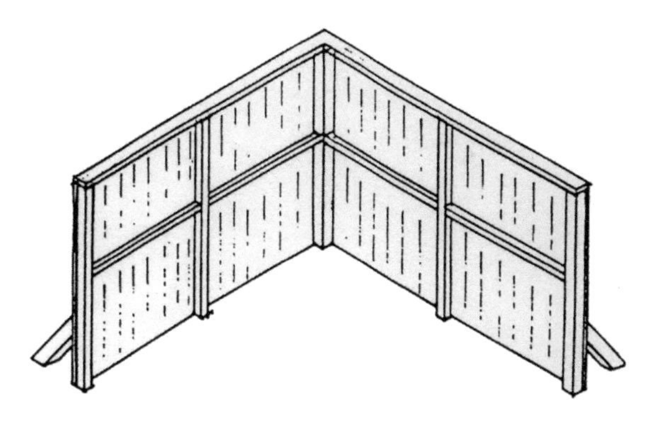

Angled shelter

8.1 Methods of providing shelter

Poisonous plants

All paddocks used for horses must be kept free of weeds and poisonous plants as they take up valuable grass space and are harmful to the horse if eaten.

Any poisonous plants found in the field should be completely uprooted, removed from the field and burnt. Large trees and hedges will need to be well fenced off, ensuring that no leaves or berries can fall or be blown within the horses' reach.

Poisonous plants that are common in the UK include:

Ragwort. This has yellow, daisy-like flowers and is deadly poisonous. Horses will not usually eat it while it is growing but it becomes more palatable once it has wilted. Ragwort is particularly lethal when eaten in hay.

Laurel, privet, yew, box, rhododendron and laburnum. These are usually found in gardens – be sure that if any of these appear in gardens neighbouring the field they are not within the horses' reach. It is also worth explaining to the owners of the properties that clippings should not be thrown into the field. (Some people might do this unwittingly, thinking they are giving the horses a treat.)

Foxglove. Horses do not usually eat fresh foxgloves, although they become more palatable in hay.

Acorns. Horses should not be allowed to eat fallen acorns – shovel up the acorns and keep the horses out of that field until they have finished falling.

Buttercups. Fresh buttercups are poisonous but they are harmless if eaten in hay.

Bracken. Horses will normally avoid eating bracken – however, on some commons and hills it is so abundant that ingestion cannot be avoided.

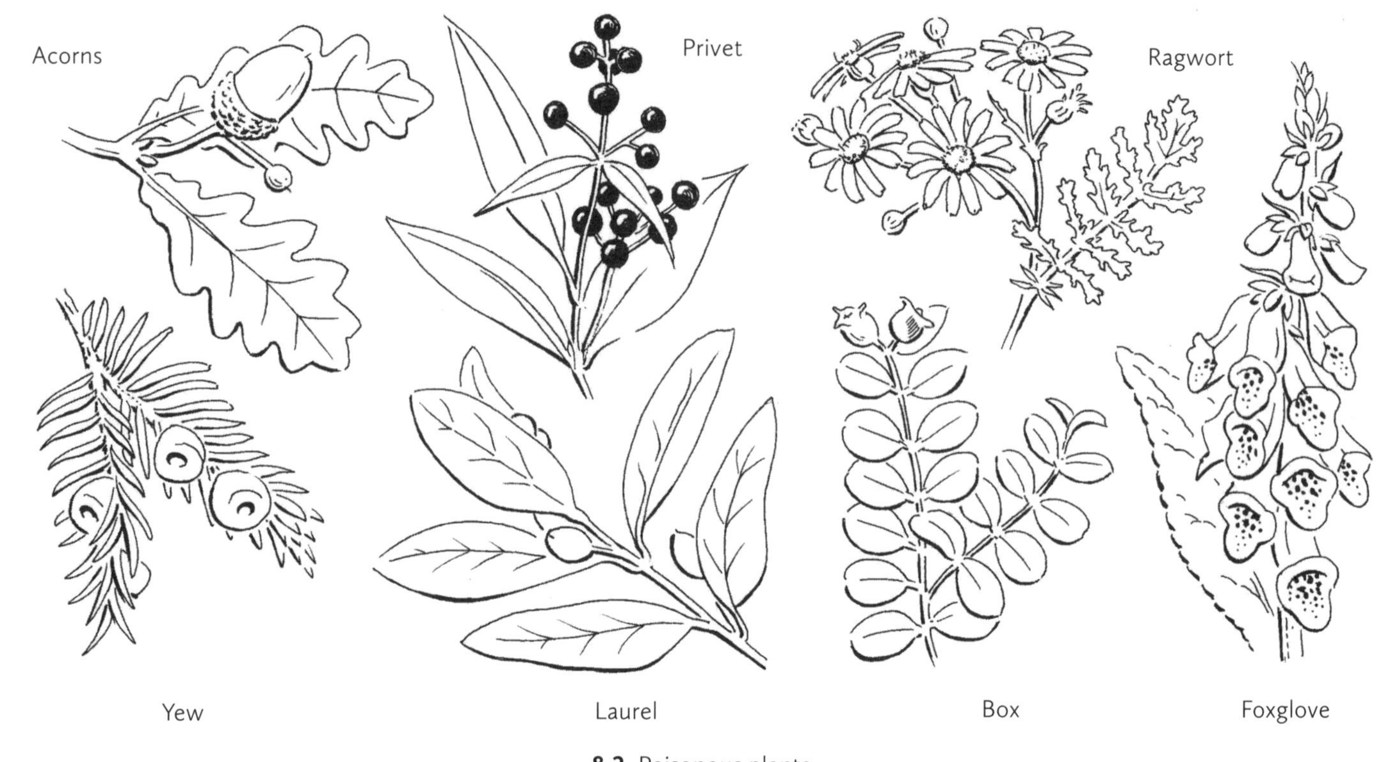

8.2 Poisonous plants

ITQ 8.3

Why is barbed wire unsuitable for use as fencing for horses?

ITQ 8.4

Give two uses of electric fencing:

1.

2.

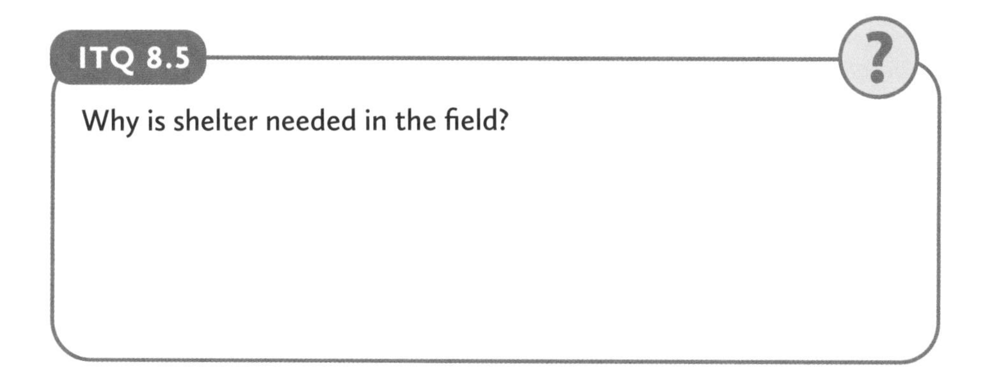

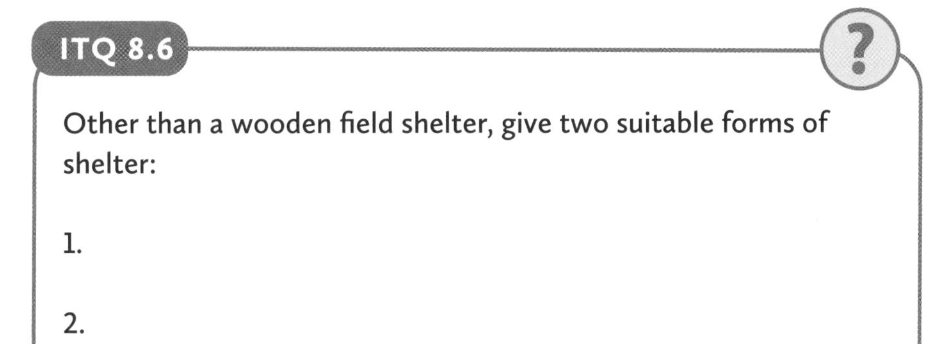

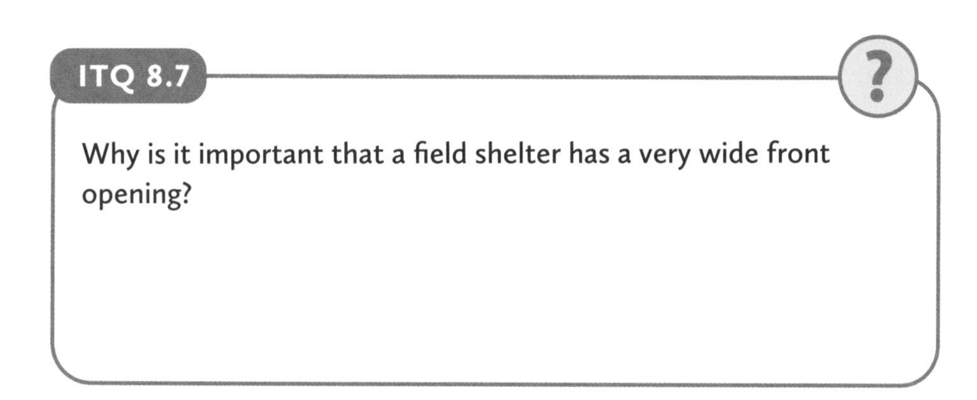

CARE OF THE PADDOCK

In order to get the best from the grazing and to ensure the horses' safety, regular paddock maintenance is necessary. Paddocks which are untidy, covered in droppings and over-grazed with patches of coarse, long grass are referred to as being **'horse- sick'**.

The over-grazed areas of a field are known as **'lawns'** and the areas where the horses do not graze are known as **'roughs'.**

Droppings

- Horses will not usually graze around droppings. This results in long, rough grass stalks growing. Collect the droppings as regularly as possible – ideally every day.

Worm larvae are passed out of the horse's body in the droppings, and then move onto the grass stalks. Eventually the worm larvae are eaten, either by the same or another horse, develop into mature adult worms and lay more eggs which develop into larvae and are passed out in the droppings onto the field. This is how many internal parasites continue their life-cycle.

- Regular picking up of the droppings can help to reduce the level of infestation. *All horses need to be wormed regularly* (the frequency depends on the type of drug used and the pasture management regime) to kill any worms and larvae that are present in their digestive tract.

- In the hot, dry weather the fields could be harrowed to spread the droppings out and dry out and kill the larvae. This is not as satisfactory as removing the droppings but may be more practical on a large area. Do not harrow in warm, wet weather as this spreads the larvae in ideal conditions for their development.

Field rotation and maintenance

- Fields should be divided into at least two smaller paddocks, to allow for rest and rotation. Fields will become horse-sick if over-grazed.

- Sheep or cattle can graze the paddock as they eat the longer, rough stalks that horses are not fond of. Worm larvae are destroyed in the stomachs of a cow or sheep, helping to control infestation.

- **Topping** the grass (cutting the top few inches off with a grass cutter) promotes more vigorous growth and removes the long, rough patches. It also removes the top stalks to which the worm larvae are attached.

- The grass should be given time to rest and recover each year. The land will also need to be harrowed, fertilised and limed to promote good grass growth – a local farmer may be able to advise on this.

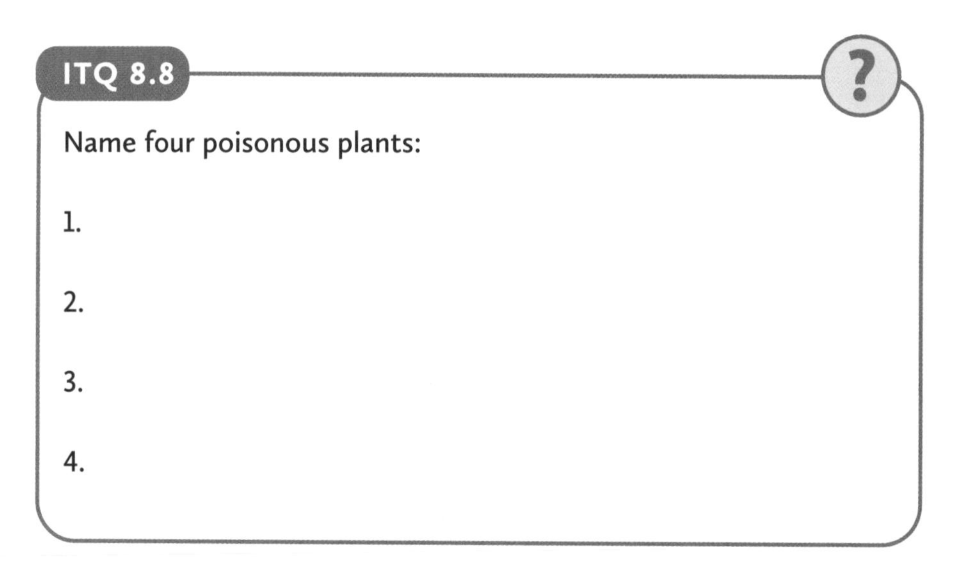

ITQ 8.8

?

Name four poisonous plants:

1.

2.

3.

4.

ITQ 8.9

What is meant by the term 'horse-sick pasture'?

ITQ 8.10

Why is it important to collect droppings from the pasture regularly?

ITQ 8.11

a. In a horse-sick paddock, what is a lawn?

b. In a horse-sick paddock, what is a rough?

ITQ 8.12

a. What should you check for each day in the horses' field?

b. Why is it important to check the field daily?

CHECKING HORSES AT GRASS

Vigilance

Horses out at grass must be checked at least twice a day for injuries, illness or lameness. If the field is not close to your home, ask a nearby neighbour to watch out for anything out of the ordinary. Give them your telephone number so that you can be contacted if needed.

All horses and ponies should be freeze-marked and/or micro-chipped. Freeze-marking is a painless form of branding and helps to deter thieves from stealing the animals. As freeze-marked horses are on a central register, they can be traced by the police more easily. In micro-chipping, micro-chips are inserted into the large ligament of the neck and can be detected using a scanner.

Seasonal care

When caring for horses at grass the different seasons bring their own particular problems that you must be aware of.

Spring

Feeding

Depending on the weather, the grass normally grows quickly in the spring and can be very rich and lush. Some horses get fat and will need to be kept partially stabled. An alternative to stabling is to provide a 'starvation' patch – a small fenced-off area of the field which the horses graze down.

Health

If a horse has access to a lot of grass, there is a risk that he will eat too much which can lead to **laminitis**, an ailment which affects the blood supply and tissues within the hooves, causing great pain and lameness. Therefore, horses or ponies prone to laminitis will need to have access to grass restricted.

Another ailment that affects horses in the spring is **sweet itch**. This is an allergy to the saliva of biting midges. Horses suffering from sweet itch cannot be outside at dawn or dusk, so can only go out for a few hours during the day.

If it is a very wet spring, the bacteria that cause **mud fever** will be prevalent. It is important to check the horse's heels for signs of inflammation and soreness.

Summer

Feeding

During a very dry summer the grass may not grow well – attention must be paid to the horses' condition as they will not gain enough from the grass and may need additional feed and/or hay. If, on the other hand, grass is plentiful the horses may get fat and need their access to grass reduced.

Heat and flies

Flies can be irritating to horses in the summer, often swarming around their eyes and causing inflammation. Horseflies are particularly annoying as they bite. Various fly repellents are available, with varying levels of success. A fly fringe can be attached to the headcollar (leather headcollar only, for safety) to keep the flies out of the eyes.

Most horses do not enjoy intense heat and are often happier to be stabled during the hottest part of the day.

Watering

The water trough must be checked frequently as horses will drink more in hot weather.

Health

Laminitis and sweet itch are still a problem and so special care must be taken with horses known to suffer from these ailments.

Autumn

Feeding

Depending on the weather, the grass is not normally as abundant at this time and starts to lose its feed value. Later in the autumn you must start feeding hay to supplement the diet and prevent the horses from losing condition.

Watering

Watch that fallen leaves do not block and foul the water trough.

Warmth

As it starts to get colder at nights, thin-skinned horses may need to be rugged up.

Winter

Feeding

By the time winter arrives the grass has no feed value – for most of the winter it may be frozen and covered in snow or completely poached. Your horse must receive plenty of hay and, if you are riding him regularly, short feed as well, e.g. horse and pony cubes, sugar beet pulp and chaff.

When feeding horses in the field, the following must be considered:

- If you have more than one horse in a field, keep them well apart at feeding time to prevent bullying and fighting. It may be necessary to remove either the most timid horse or the bully to ensure that each receives the correct amount of food.
- Feed bowls should be of a design that cannot be kicked over.
- Always put out more piles of hay than the number of horses – this way every horse should get sufficient ration.
- Haynets hang lower when empty – this can be dangerous as a horse may get a foot caught. For this reason haynets are best avoided in fields – horses who play-fight can get caught up, even in nets that are safely tied up.

Watering

In the winter the water troughs and pipes will freeze so it is important that you check these three times a day, break the ice and remove it from the trough, or carry water buckets to the field.

Warmth
Check that the horses are warm enough. A cold horse will use energy to keep warm, resulting in a loss of condition. Signs that a horse is cold include:

- Shivering.
- Staring coat (hairs stand on end and the coat looks dull).
- Tips of the ears feel cold.
- May appear miserable – standing in the field with the head hanging low.

To warm a horse up:

- Put on a thick, dry turnout rug. A hood may also be needed.
- Give a high-carbohydrate feed.
- Feed plenty of hay. Hay contains **cellulose**, a form of carbohydrate, which provides slow-release energy and is therefore warming.
- Exercise will generate body heat through muscular activity.
- If the horse is extremely cold it will be necessary to bring him into a stable, rug him up and put woollen stable bandages on. Keep the top door open to ensure adequate fresh air.

If the horse wears a turnout rug you must check the following:

- The rug must keep the horse dry. Slide your hand beneath the rug and check that it is still waterproof. If the rug is leaking or is thoroughly soaked, it must be replaced with a dry one. All turnout rugs need regular re-proofing with a water repellent spray or cream.
- The rug must fit properly and not chafe or rub. Turnout rugs are inclined to slip back, which can lead to rubbing across the shoulders.
- Use the correct weight of rug according to the weather conditions. In severe weather a neck hood can help to protect the horse against the elements.

Health
Problems which occur in winter include **mud fever** and **cracked heels.** Both ailments can be caused by standing constantly in wet, muddy conditions.

Turning out and catching

Turning out

Leading horses to the field and turning them out can be dangerous, particularly if the horses are young and/or excitable.

The correct and safe method of turning out several horses is as follows:

- If well-mannered, the horse can be led in a headcollar to the field. Inexperienced handlers should not turn out difficult horses.
- Gloves should always be worn when leading a horse to prevent rope burns.
- Lead the horse(s) into the field and close the gate properly.
- Lead them a little away from the fence and turn them so they are facing the gate.
- Make the horses stand and check that everyone else is ready before unfastening the buckles of the headcollar.
- Everyone must undo and remove headcollars at the same time, stepping backwards as they do so. This prevents the handlers from getting kicked should the horses turn and gallop off, kicking and bucking.
- If a horse has been stabled for a long time he will be more inclined to buck and kick when he is turned out. You must be especially careful to avoid getting kicked.
- If a horse is turned out on his own and is not used to this, he may get quite upset. The horse may pace up and down the fence line, calling to his companions. In extreme circumstances the horse may jump out of the field. Although many horses are quite happy on their own, it is better for the horse to have a companion as horses are naturally herd animals.

Catching

When catching a horse you will need a headcollar, lead-rope and possibly a small titbit, e.g. a couple of horse and pony cubes or a slice of apple or carrot.

It is not safe to take a bucket of food into a field if there is more than one horse, as you will attract the others who will group around you, trying to get at the feed. They normally then start fighting amongst themselves and you could get kicked or trampled.

A headcollar should, ideally, not be left on a horse whilst he is turned out because of the danger of him catching it on a low branch or other projection. If a horse is very hard to catch he may be left with a *leather* headcollar on – *never* a nylon one. In the event of the headcollar being caught up, the leather one would break, whereas a nylon one would not, which may result in the horse injuring himself.

Always speak when approaching the horse in the field, especially if he is not facing you. This ensures that he knows you are there and avoids giving him a fright. Ideally, approach a horse from an angle (e.g. towards his shoulder) whereby he can see you.

Approach quietly at his nearside (left) shoulder, talking gently as you do so. If you run up to him, arms flapping and shouting, you will give him the horrors and he will probably gallop off to the other end of the field.

As you reach him, offer him the titbit on your flat palm and, as he eats it, slip the lead-rope over his neck, so that you have some control. Then slide the noseband of the headcollar over his nose, pass the headpiece over the top of his head behind his ears and fasten.

Make sure that, as you lead him out of the field, none of the others can escape as you go through the gate.

ITQ 8.13

Give three important points to remember when turning horses out into the paddock:

1.

2.

3.

ITQ 8.14

Give three important points to remember when catching horses out in the paddock:

1.

2.

3.

9 Feeding and Watering

REQUIRED SKILLS/KNOWLEDGE	Learnt, revised, practised?	Confirmed
How to feed and water horses.		
• Know the rules of feeding and watering horses.	☐	☐
• Be familiar with feedstuffs and forage in general use.	☐	☐
• Understand the importance of hygiene and of feeding quality foodstuffs.	☐	☐
• Methods of feeding and watering horses kept at grass and/or stabled.	☐	☐
• Feeding regimes for grass-kept and stabled horses.	☐	☐

FEEDSTUFFS

General categories

Feedstuffs may be divided into the following categories:

Cereals (also known as 'straights' or 'concentrates'). These include grain such as oats, barley and maize. The cereals are often prepared in such a way as to improve digestibility. This includes boiling, rolling, bruising, flaking or a heat treatment called **micronising**.

Protein feeds. These are either of animal origin such as dried milk, or of plant origin such as beans, peas and linseed.

Bulk feeds. Bran, sugar beet pulp, grass meal and chaff add bulk (roughage/fibre) to the diet and aid digestion.

Compound feeds (together with cereals these are also know as **'concentrates'**). These include the extensive and ever-expanding range of cubes and coarse mixes. They are prepared to include the necessary balance of nutrients for each specific area, e.g. resting horses, competition horses, breeding stock, etc.

Forages. These include grass, hay and haylage. In addition to the bulk feeds, forages provide fibre (also known as roughage) which is essential for healthy gut function. The fibre is provided in the form of **cellulose** – a fermentable carbohydrate that the horse digests in his hind gut.

Cereals

Cereals, by comparison with other feeds we give our horses, are high in quick-release energy sources which include sugars and starch. These energy sources can have the effect of making a horse excitable – some people refer to this as the 'heating effect'. Many people believe that oats are the main culprit for causing horses to be excitable but they actually have a lower energy level weight for weight than any of the other cereals. Oats are also the grain which is lowest in starch, but the starch in them is said to be more readily digestible than that contained in other cereals, so this might be partly responsible for the 'exciting' effect that oats have on some horses.

As a rule of thumb, cereals need to be fed to horses using a considered approach; the majority of horses in medium work could easily be fed a diet free from cereals and still have plenty of energy to carry out the work required of them. Cereals can form a useful part of the diet of a working horse provided that they are not fed in too large a quantity as this is likely to cause behavioural issues in some horses.

9.1 Rolled oats

Oats

- For the reasons outlined above, oats are best fed to horses in hard work only.
- Oats are normally **rolled** or **bruised** to aid digestibility – this is achieved as the processing breaks the hard, indigestible outer husk. Oats can also be fed in their **naked** form, meaning that the outer husk is not present and the energy in them is much more available to the horse.

9.2 Whole barley

Barley

- Barley can be bought cooked and flaked, or micronised.
- This is a fattening feedstuff as it has a high carbohydrate content and excess carbohydrate will be stored by the horse as fat.
- Whole uncooked barley may be boiled and fed.

9.3 Micronised barley

EXAM TIP

Make sure you can tell the difference between oats and barley; oats are longer and slimmer, barley more rounded.

Maize

- Because of its high carbohydrate content, maize is unsuitable for many horses and ponies unless they are in very hard work.
- Maize, like barley, is a fattening feedstuff.
- It is fed cooked and flaked or micronised.

9.4 Whole maize

Protein feeds

Beans and peas

- These may be fed crushed, split, steamed and flaked or micronised to a horse requiring increased protein levels and are commonly added to coarse mixes.

Linseed

- Linseed (the seed of the flax plant) is very high in oils and protein.
- These small flat, brown seeds are poisonous to horses in their raw form and so must be cooked and processed before being fed. Linseed can be fed to horses in the form of linseed jelly, tea, cake or oil.

9.5 Flaked maize

Bulk feeds

Bran

- Bran is the by-product of wheat milling and has a low energy value. Because the milling process nowadays is far more efficient than it was years ago, bran is now commonly dustier than it used to be and contains flakes that are much smaller.
- It contains a high level of fibre.
- Bran should always be fed dampened, as dry bran can cause the horse to choke.
- It is not suitable to feed in large quantities as it contains too much phosphorous and not enough calcium. The effect of this leads to impaired bone growth which could cause significant problems in the young, growing horse.

9.6 Bran

EXAM TIP

Whilst you need to know about bran, it is not commonly fed – avoid mentioning it if talking about suitable feedstuffs in your exam.

9.7 Sugar beet cubes

9.8 Sugar beet flakes

9.9 Soaked sugar beet

Sugar beet pulp

- Sugar beet comes in dried form and is either shredded or in cubes.
- On no account must it be fed dry. It must always be well soaked – shredded beet for at least 12 hours, beet cubes for at least 24 hours. This is to prevent the pulp swelling up inside the horse's digestive tract and causing colic. However, it is now possible to buy sugar beet treated so that it only requires soaking for 10 minutes and this can be very convenient, especially if one forgets to soak the shreds or cubes in time for the next feed!
- Pulp is a good fibre feed and is non-heating when fed in its unmolassed form so is therefore safe to feed to all horses.
- Horses with a higher energy requirement can be fed molassed sugar beet pulp, which contains more sugar and therefore provides more energy than the same amount of unmolassed pulp.

EXAM TIP

Make sure you can tell the difference between sugar beet cubes and horse and pony cubes. Sugar beet cubes are usually smaller and darker than other types of cubes.

Grass meal

- This is non-heating and often comes in the form of pellets; it is included in many coarse mixes.
- Grass meal is nutritious and useful to feed in winter when there is no feed value in the grass.

Chaff

- The old-fashioned type of chaff was simply chopped hay and straw.
- Chaff can now be purchased with additives such as molasses, limestone flour, garlic, vitamins and minerals.
- Feeding alfalfa chaff is now very popular. Alfalfa is a legume crop which provides higher protein levels than other types and is nutritionally a better 'mixer' than bran. It aids digestion and helps to prevent a horse bolting his food.
- Unmolassed varieties can be fed safely to all horses and ponies dampened in with the short feed.
- Chaff can be added to the feed to 'bulk' it out and satisfy the horse's appetite. This is useful when the horse receives only a small ration.

Compound feeds

9.10 Cubes

Cubes

There is a range of different types of cube, all very similar in appearance, hence it is difficult to identify individual types of cubes. The information on the feed sack will describe the content of each type of cube.

Water must always be available to a horse eating cubes, as they are dry. This is an important point to remember as choking and colic may result if water is not available.

Complete cubes

- Complete cubes are very high in fibre as they contain the forage ration as well as the short feed.
- Complete cubes may be fed to replace the hay ration if a horse suffers from a very bad dust allergy, although it is possible to buy dust-extracted, vacuum-packed hay or haylage which is more interesting for the horse.

Horse and pony cubes

- These have a good balance of all the necessary vitamins, minerals, proteins and carbohydrates that a horse in light to medium work requires.
- They are specially designed to be non-heating, so may be fed safely to most horses or ponies.
- It is quite safe to feed nothing but horse and pony cubes, although some horses may become bored and go off them.

Stud cubes

- Stud cubes are high-protein cubes designed to meet the needs of broodmares, their foals, stallions and youngstock.

Racehorse/event cubes

- These are high in carbohydrates and meet the energy needs of high-performance competition horses.

Coarse mixes

9.11 Coarse mix

- These comprise a highly palatable mixture of cereals and bulk feeds with added vitamins and minerals.
- High-protein coarse mixes will contain split beans, peas and protein cubes in addition to the cubes, barley, oats, grass meal, molasses and linseed cake contained in the lower protein mixtures.
- Coarse mixes are designed to make up the whole concentrate ration to be fed with the normal quantities of hay. It is quite common practice to mix them with other feedstuffs such as chaff and pulp but it must be borne in mind that doing this might unbalance an otherwise balanced concentrate ration and so is probably best avoided.

- As with cubes, coarse mixes are prepared to meet the differing needs of all types of horse and pony. There is a complete range starting with non-heating, low-energy mixes, through to high-energy competition mixes.

Quality of concentrates

Only food of good quality should be fed to horses. All feed must be stored in a dry vermin-proof environment, e.g. in plastic or metal bins in the feed room, and used within its 'use-by' date. Instead of appearing dry and fresh-smelling, feed that has been incorrectly stored, i.e. allowed to get damp and/or soiled by vermin, may appear:

- Mouldy – clumped and coated with white or green mould.
- Stale – smell 'musty.'
- Dusty – may disintegrate into very small particles.

Food matching any of the above descriptions should not be fed to horses.

EXAM TIP

Practise identifying the various feedstuffs present in coarse mixes.

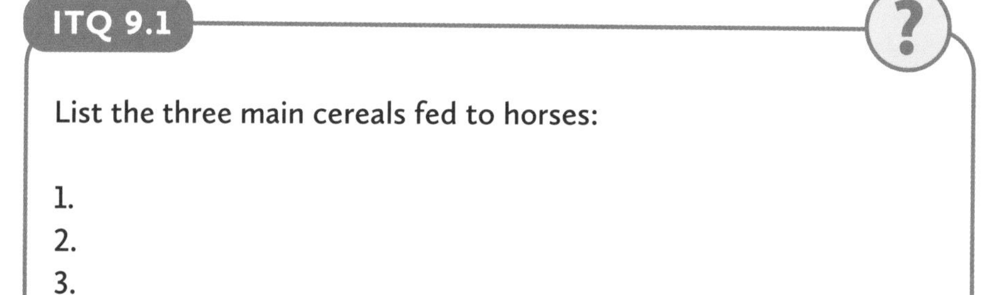

ITQ 9.1

List the three main cereals fed to horses:

1.
2.
3.

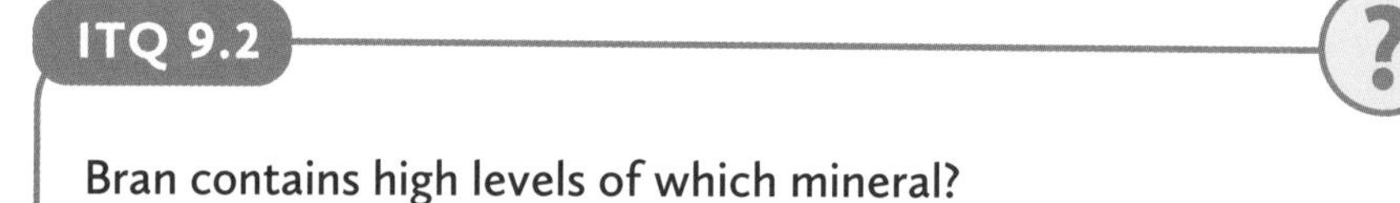

ITQ 9.2

Bran contains high levels of which mineral?

ITQ 9.3

How should dried sugar beet pulp cubes be prepared before being fed?

ITQ 9.4

What are the advantages of feeding chaff?

IN-TEXT ACTIVITY

Using magazines/advertising/feed suppliers as sources of information, select four feed manufacturers and obtain the following information:

Manufacturer's name	Name of non-heating cubes	Cost per 20kg bag	Name of high energy/competition cubes	Cost per 20kg bag
1.				
2.				
3.				
4.				

Forages

We have discussed the feedstuffs which make up the 'short feed' or 'concentrate' part of the horse's daily ration. Hay, or haylage, should make up the larger part of the ration, especially when the horse spends long periods stabled.

Hay

Hay, the most important feedstuff for horses, is grass that has been cut and dried. It varies in quality, depending on the types of grasses, the weather when the hay lay on the fields, its age and how well it has been stored.

Meadow hay is taken from pasture normally used for grazing and is generally softer than **seed hay**, which is a specially sown crop containing fewer grass varieties.

Poor-quality hay should never be fed, but even good hay contains a significant amount of dust. Some horses are allergic to the dust found in hay. The dust allergy affects the respiratory system and causes the horse to cough. With these horses it will be necessary to soak the hay for approximately 20 minutes before feeding. This makes the dust particles stick to the hay, which then prevents the dust from being inhaled.

Hay should not be soaked overnight as the nutrients are washed out, reducing the hay's feed value. Poor-quality, dusty hay should never be fed – soaking will do nothing to enhance its feed value or palatability. The only circumstance in which you might consider feeding hay that has been soaked for significantly longer than 20 minutes is if you are feeding a very overweight or laminitic horse or pony and you want them to eat a hay ration that contains few nutrients. In these cases soaking hay for long periods (whilst being careful to change the soak water regularly) can be very useful and will keep the horse or pony occupied and happy without overfeeding nutrients.

If a horse is *very* allergic to hay, he will need to be fed **haylage** or **vacuum-packed (bagged) hay**. These are dust-free alternatives to normal hay. They are expensive compared to normal hay.

Qualities of good hay

- The first thing you will notice is the way the hay smells. It should have a good **'nose'**. This means it should smell sweet and pleasant, never musty.
- The bales must be dry – if they have been stacked whilst damp they will soon start to rot.
- It must not be dusty or mouldy – if mouldy hay is fed the horse will soon start coughing as he will inhale mould and dust spores. Mouldy hay can also cause colic.
- When you cut the strings the bale should fall loosely apart.
- The hay should be greenish-light brown in colour, depending on the grasses used. Meadow hay is normally greener than seed hay.
- There should be a good selection of grasses in a bale of meadow hay.
- It must be completely free from poisonous plants, especially ragwort. Buttercups are harmless in hay.
- The hay must be free from weeds. A few thistles will do no harm, but too many should be avoided.

Qualities of good haylage

Haylage is grass cut between heading and flowering and left to dry partially before being baled and wrapped in plastic.

- The bag must not be punctured as this causes the haylage to rot.
- Haylage smells different from hay because of the fermentation process: the haylage must smell clean, not unpleasantly pungent or over-fermented.
- It must be free of mould: mouldy haylage can cause respiratory problems and colic.

Points to remember when storing hay

- Hay is expensive. It must always be stored in a dry place to avoid waste.
- Hay may be fed from a haynet to prevent wastage, although haynets in a field can be hazardous.
- Feeding hay from the floor is a natural position from which the horse can eat.

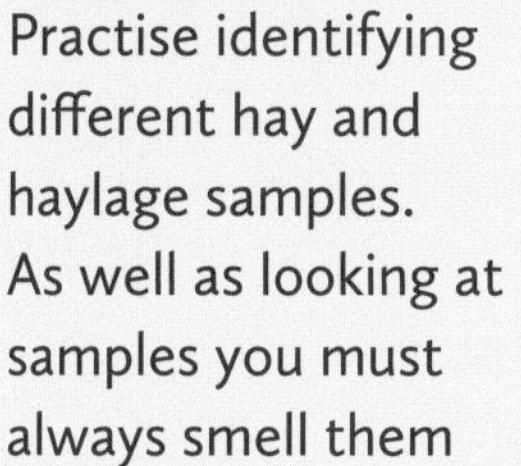
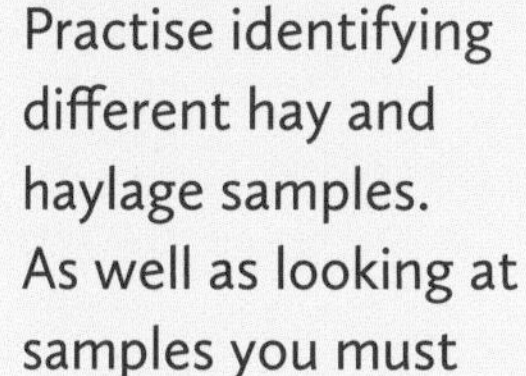

EXAM TIP

Practise identifying different hay and haylage samples.
As well as looking at samples you must always smell them too.

ITQ 9.5

What is the difference between meadow hay and seed hay?

ITQ 9.6

List three qualities of good hay:

1.

2.

3.

THE RULES OF FEEDING

When we talk about feeding horses there are certain well-established rules that must be followed. These rules are known as **'The Rules of Feeding'** and must be learnt by everyone involved with horses. Using these rules as the basis for feeding horses will help keep the horse in good health and help you avoid making mistakes which could lead to colic or other problems associated with diet.

1. **Feed little and often.** Relative to his size, a horse has a small stomach. It is approximately the size of a rugby ball and has a capacity of between 9 and 12 litres (about 16–21 pints). In the wild a horse grazes on and off all day so that his stomach is always about two-thirds full; the horse's stomach is most efficient if it remains about two-thirds full all of the time.

It is, therefore, more natural to give the horse two or three smaller feeds a day as well as a constant supply of forage, rather than one large concentrate feed and two or three haynets, especially if the horse is stabled.

2. **Feed at regular times each day.** Horses are creatures of habit and soon learn when to expect their next meal. Each day you should try to feed at more or less the same time so a routine is developed.

3. **Feed plenty of bulk (roughage/fibre).** As a horse's natural diet is grass, his digestive system is designed to ferment and digest large quantities of roughage. Hay is dried grass so provides the necessary roughage, helping the digestive system to work properly. Eating hay also helps to alleviate boredom in the stabled horse and the action of chewing stimulates the horse to produce saliva which helps to prevent problems further down the digestive tract.

4. **Feed only good quality food and hay.** All food should be fresh, free of mould and/or dust and have a pleasant smell. Some feedstuffs can be stored longer than others, e.g. horse and pony cubes last well, but bran absorbs moisture from the atmosphere so does not keep for long. As mentioned, dusty or mouldy hay will cause health problems in horses and should always be thrown away.

5. **If water is not always available, water before feeding.** Horses should have clean, fresh water available to them at all times. If however, you have been riding, e.g. at a show or schooling, you will need to offer the horse a drink well before you feed. This is to prevent him taking a long drink immediately after eating, which could wash most of the food through his stomach before the digestive process started.

6. **Introduce changes gradually.** Horses get used to particular types of food and if their diet is suddenly changed they may not be able to digest the new food properly.

   In the event of a sudden change there may not be the correct conditions in the horse's digestive tract to digest the new feedstuff. Therefore, always try to add a little of the new food at a time, gradually building up the amounts given. Sudden changes in feedstuff and grazing quality can lead to colic.

7. **Allow 1½ hours after a feed before exercising.** Once a horse has eaten he requires approximately 1½ hours for the food to leave his stomach. If he is exercised strenuously soon after eating, it could cause colic as the circulatory system (heart and blood vessels) directs blood away from the digestive tract to the muscles needed for work.

   It is quite safe to ride your horse *before* he has been fed. If you ride first, allow him to cool off and have a drink before he eats.

8. **Feed something succulent every day.** If your horse lives out at grass he will be eating something succulent every day – grass. If he is stabled most of the time, or it is winter and there is no grass, you should feed something succulent such as carrots, apples or parsnips.

These must be **sliced** – never diced or cubed – as a horse could easily choke on a square piece of carrot or apple. Succulents aid digestion and add variety to the diet.

9 **Always keep feeding equipment clean.** Feed scoops, buckets and mangers must be scrubbed and rinsed out regularly in order to keep them clean. Bad hygiene can lead to rotten food building up, which not only smells bad but can cause harmful bacteria to develop.

10 **Feed according to work, type, age, size, temperament and time of year.** This is probably one of the most important rules of feeding so we will break it down further into the different headings below as each one is important. More harm is caused through overfeeding than any other type of mismanagement problem. Overfeeding can make the horse fat, badly behaved and cause laminitis.

Feed according to workload. If a horse is not working he will require food purely to keep him alive and looking well. This is referred to as **maintenance rations**. In the summer, good grass will provide the maintenance ration. If the horse does not have access to good grass, lots of good-quality meadow hay should be sufficient.

A horse in work will need extra energy – if a working horse is not given extra food he may lose weight. However, if he is not working very hard, he will not require lots of extra food.

Feed according to type and temperament. An excitable horse will need non-heating (low-carbohydrate) food, whereas a lazy cob type may benefit from higher-energy foods.

Thoroughbreds tend to need proportionately more food to keep their condition than natives. However, some Thoroughbreds can be highly strung and excitable so need non-heating feeds.

Small ponies do not cope with heating feedstuffs at all well. They become too 'fizzy' for their young riders and are very prone to developing laminitis.

Feed according to age

A young, growing horse up to the age of 4 years needs a balance of vitamins, minerals and protein to ensure good bone and tissue growth.

An old horse of, say, 20 years or more may need plenty of bulk (hay) and quite fattening foods to prevent him from losing weight. Many of the manufacturers produce mixes specially designed to meet the nutritional needs of the older horse.

Feed according to size and bodyweight

Obviously you would not feed a small pony the same amount as a big horse. Usually the smaller the pony, the less he has to eat, depending on his type, temperament and work done. The amount to feed is calculated according to the weight of the horse or pony.

Within this book we will not be looking at detailed ration calculation, but we'll start to look at some samples of daily rations to give you the idea of how much food a horse or pony needs in a day (see next heading).

Feed according to the time of year

This applies particularly to horses who spend time out at grass. The grass has no feed value during the winter months so additional hay will be needed. Grass-kept horses need extra hay and food in winter to help keep them warm – the process of digesting hay generates body heat.

Once the grass starts to grow in the spring no extra hay will be needed whilst the horse is in the field. If the grass is very lush and/or the horse is prone to laminitis, it will be necessary to restrict his access to the grass either by stabling him or dividing the field.

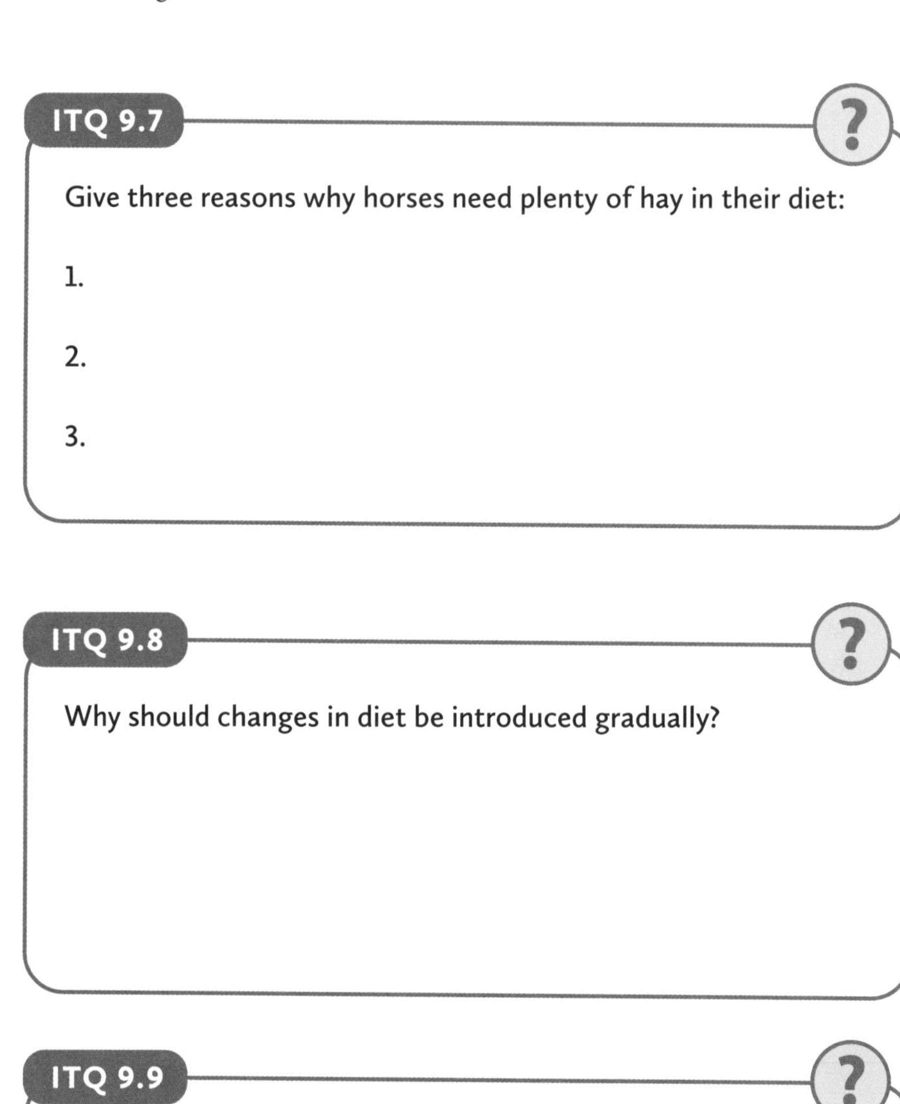

ITQ 9.7

Give three reasons why horses need plenty of hay in their diet:

1.

2.

3.

ITQ 9.8

Why should changes in diet be introduced gradually?

ITQ 9.9

What is meant by 'maintenance rations'?

Sample feed rations

Useful measurements:

1kg = 2.2lb
1lb = 0.45kg

The average section of hay weighs approximately 1.8kg (4lb).
When filled level, a normal 1,500ml (2½pint) feed scoop holds the following:

1.36kg (3lb) cubes
0.56kg (1¼lb) flaked barley
1.36kg (3lb) soaked sugar beet pulp
0.33kg (0.75lb) chaff
0.69kg (1½lb) coarse mix
1.36kg (3lb) carrots
0.45kg (1lb) bran

The following feed charts have been devised for horses and ponies turned out all day and stabled at night. They are in light work – in your exam light work is defined as 'daily walk, trot, canter where the horse is not stressed'.

Assume that they are all around 8 years old, of calm disposition and maintain condition well. There are many types of feedstuffs to choose from – while those used in the following sample rations are suitable there may be other similar feeds which are equally suitable.

All weights shown are **dry weights** and whilst many feeds may be fed dry, it improves mastication and aids digestion if the feeds are well dampened. Many owners like to feed sugar beet pulp as it dampens the feed, provides a non-heating energy and fibre source, keeps weight on and is suitable for all horses and ponies.

When calculating the ration to include sugar beet, weigh the beet *dry* (unsoaked). Do not take into account the weight of the water in which the pulp has been soaked or cooked, as the water does not contain any nutrients – these are derived from the dry feed.

Ponies and hardier types of horses are much less likely to require large amounts of feed as they have evolved to live in harsher environments and on less food than finer horses – this should always be borne in mind when planning diets as overfeeding can be very dangerous.

The examples that follow are based on ponies and horses who are being fed a set amount of hay and concentrate feeds. Rations are often worked out in these terms and are based on a forage: concentrate ratio – in simple terms the more work the horse or pony is doing, the higher the percentage of concentrates he can safely receive. Horses and ponies in light work or on a maintenance ration can often remain very healthy and looking well eating only fibre. The percentage of concentrate feed in a ration should never be greater than 65 per cent.

It is becoming more common for nutritionists and vets to recommend feeding horses *ad lib* forage and disregarding the older system of using a forage: concentrate ratio. This can be very useful for the vast majority of horses as it is much more harmonious with their bodies and mimics the way they would feed in the wild. Feeding forage in this way can also save money in the long run as it is likely that the horse will need less expensive hard feed if he has access to a constant supply of good-quality forage. However, in your exam you will be required to estimate total quantity and therefore it is useful for us to provide examples which include a definitive amount of forage. This could be split equally into two feeds:

Stabled 13hh pony

13hh pony	Metric	Imperial
Concentrates Hay	0.9kg 3.6kg	2lb 8lb
Daily total	**4.5kg**	**10lb**

This could be split equally into two feeds:

	MORNING AND EVENING FEEDS Metric	 Imperial
Horse and pony cubes	0.22kg	½lb
Chaff	0.115kg	¼lb
*Sugar beet	0.115kg	¼lb
Amount per feed	0.45kg	1lb
Daily total of 2 feeds	**0.9kg**	**2lb**

With hay fed as follows:

Morning hay	–	–
Evening hay	3.6kg	8lb
Daily total hay	**3.6kg**	**8lb**
Total daily amount fed	**4.5kg**	**10lb**

Important points to remember

- *The weight of the sugar beet shown is the *dry* weight – you do not include the weight of water in your calculations because water, although essential for life, does not contain a significant amount of nutrients. (Water contains a few trace minerals but no protein, carbohydrates or lipids). Sugar beet *must always be thoroughly soaked* before feeding.

- In the stabled pony example ration, the pony has not been given any hay in the morning because he has been turned out to grass during the day. If there was snow on the ground hay could be given in the morning and slightly more hay given at night.

- If the pony was kept constantly in the stable (not recommended) the ration would be split into three feeds a day.

- In summer this pony's ration would be reduced according to the quality of the grass he has access to whilst turned out. If constantly stabled the ration would remain unaltered. A 13hh pony should be quite able to carry out light work with no extra feed.

- Remember to work either in kilos (kg) or pounds (lb) – don't mix the two or you will get very confused! When you add up the columns to check the totals, make sure you add *either* the measurements shown in kilos or pounds – don't add up a mixture of the two. Also, bear in mind that, because you are using approximate (rather than absolute) equivalents between the two systems, there will be times when comparative tables and charts will show minor anomalies. While these are inevitable, try to keep any calculations you make as accurate as practical.

- Never feed ponies diets that are high in cereals or high-energy coarse mixes as they can cause laminitis, obesity and bad behaviour.

14hh pony

- Late autumn/winter.

- Light work so a ratio of 80 per cent forage to 20 per cent concentrates has been used.

- Good doer – keeps condition well.

- Stabled at night and out at grass all day – no feed value in grass.

14hh pony	Metric	Imperial
Concentrates Hay	1.36kg 6.8kg	3lb 15lb
Daily total	**8.16kg**	**18lb**

	MORNING AND EVENING FEEDS Metric	Imperial
Horse and pony cubes	0.45kg	1lb
Chaff	0.115kg	¼lb
*Sugar beet	0.115kg	¼lb
Amount per feed	0.68kg	1½lb
Daily total of 2 feeds	**1.36kg**	**3lb**

Morning hay	2.3kg	5lb
Evening hay	4.5kg	10lb
Daily total hay	**6.8kg**	**15lb**
Total daily amount fed	**8.16kg**	**18lb**

Important points to remember

- In this ration some of the hay has been put out in the field in the morning.
- If the pony needs more energy the horse and pony cubes could be replaced with coarse mix.
- Sugar beet shown as dry weight (before soaking).
- In summer this pony's ration would be reduced according to the quality of the grass he has access to whilst turned out. If constantly stabled the ration would remain unaltered. A 14hh pony should be quite able to carry out light work with no extra feed.

15hh horse

- Late autumn/winter.
- Medium work so a ratio of approx. 60 per cent hay to 40 per cent concentrates.
- Good doer – keeps condition well.
- Stabled at night and out at grass all day – no feed value in grass.

15hh pony	**Metric**	**Imperial**
Concentrates Hay	4.1kg 6.8kg	9lb 15lb
Daily total	**10.9kg**	**24lb**

	MORNING AND EVENING FEEDS **Metric**	**Imperial**
Coarse mix	1.61kg	3½lb
Chaff	0.22kg	½lb
*Sugar beet	0.22kg	½lb
Amount per feed	**2.05kg**	**4½lb**
Daily total of 2 feeds	**4.1kg**	**9lb**

Morning hay	1.8kg	4lb
Evening hay	5kg	11lb
Daily total hay	**6.8kg**	**15lb**
Total daily amount fed	**10.9kg**	**24lb**

Important points to remember

- The energy level of the coarse mix can be chosen to suit the type of horse. A lazier sort might need a higher level of energy than a 'fizzy' sort of horse. If the horse is very lively it is better to replace the majority of coarse mix with more fibre in the form of either chaff and sugar beet or hay.
- Sugar beet shown as dry weight (before soaking).

Summary of the important points to consider when working out how much to feed a horse

- Time of year – is there any feed value in the grass?
- Is the horse turned out or stabled?
- What work is he doing? Hacking, eventing, showjumping, dressage or resting?
- His ability to maintain condition; some horses lose condition easily whilst others put weight on, on very little feed.
- Temperament – is he calm or 'fizzy'?
- Type: cobby or Thoroughbred? Light, medium or heavyweight?
- Age – very young or old?
- Is the horse susceptible to conditions such as laminitis?

Can you think of any more?

IN-TEXT ACTIVITY

Using your local yard, friends, feedstore, feed manufacturers, etc., collect and label small samples (30–60g/1–2oz) of the following feedstuffs:

a. A non-heating coarse mix
b. Horse and pony cubes
c. Dry sugar beet pulp nuts and/or shredded beet pulp
e. Oats
f. Barley
g. Bran

EXAM TIPS

Keep your calculations as simple as possible when planning and revising how much to feed in a ration.

Avoid talking about complicated rations containing too many different feedstuffs as you will be more likely to get yourself confused in an exam – keep it simple!

Always work to the nearest whole unit.

MULTIPLE CHOICE QUESTIONS

Tick the correct answer(s)
There may be more than one correct answer

ITQ 9.10

Fibre is needed for?
a. Good digestion
b. Replacement of body tissue
c. Energy
d. Healthy hooves

ITQ 9.11

Carbohydrate in the diet provides?
a. Bone strength
b. Amino acids
c. Energy
d. Oil

ITQ 9.12

Which of the following are bulk feeds?
a. Chaff
b. Horse and Pony cubes
c. Sugar beet pulp
d. Linseed

ITQ 9.13

Oats are rolled to?
a. Make them more palatable
b. Aid their digestion
c. Make them smell fresher
d. Make them last longer in storage

Watering and feeding at grass

Watering

A constant supply of clean, fresh water is essential. Whichever method of providing water is used, it should be checked regularly to ensure cleanliness.

Troughs

The self-filling type of trough is ideal. This must have rounded edges to prevent injury.

If a tap is used it must be enclosed within a wooden box to prevent a horse injuring himself on it. The trough must be emptied and scrubbed out regularly to prevent the inside becoming green and slimy as algae forms.

Pipes should be well lagged, although in the winter they may still freeze. In freezing conditions the ice will need to be broken and removed at least three times a day and the trough must then be filled with buckets of water.

The area around the trough may become wet and muddy in the winter, so some form of hard standing such as shingle on concrete should ideally be used

Buckets

Provided you do not have to carry the filled buckets too far, this is an easy method of providing water in the field. Each bucket may be stood in two old car tyres to stop them from being kicked over. Buckets must be scrubbed out regularly and filled with clean, fresh water daily.

Streams

Before allowing horses to drink from a stream you must check that it:

- Is not polluted.
- Has a safe approach, i.e. firm and not steep.
- Has a stony base, not a sandy one: horses can develop colic through drinking water which contains sand.

If the stream does not meet all of the above requirements, it must be safely fenced off and water provided in another way.

Ponds

The water in a pond is stagnant and so must never be used as a water supply for horses. Ponds should be fenced off as they can be dangerous – horses may get stuck in the mud and foals in particular, are prone to drowning in ponds.

10 Leading, Mounting and Adjusting Tack

REQUIRED SKILLS/KNOWLEDGE	Learnt, revised, practised?	Confirmed
Lead a saddled and bridled horse in hand * These aspects are covered in more detail in Section 5.		
• Lead a horse safely and effectively in walk and trot.	☐	☐
• Turn the horse safely and correctly when leading in hand.	☐	☐
Mount and dismount a horse correctly.		
• Check tack for safety prior to mounting.	☐	☐
• Mount and dismount a horse safely and correctly.	☐	☐
Adjust stirrups, girth and reins when mounted.		
• Alter the stirrups correctly.	☐	☐
• Check stirrup leathers hang correctly, i.e. are correctly rotated.	☐	☐
• Check and adjust the girth prior to riding away.	☐	☐
• Hold and adjust reins correctly.	☐	☐

10.1 Leading. See also Section 5.

PREPARATION FOR LEADING

Equipment notes:

You must:

Wear a correctly fastened crash cap.

Remove gloves for adjusting tack – put them in your pocket.

Put gloves back on prior to leading.

Carry a whip which must not exceed 75cm (30 in).

10.2 Safe method of leaving headcollar

If the horse is tied up, the headcollar will be on over the top of the bridle. The reins will be twisted around each other to take up any slack, with the throatlash passed through and fastened.

To ensure you have full control of the horse, before you remove the headcollar, standing on the horse's nearside:

- Undo the throatlash.
- Remove it from the reins and untwist the reins.
- Refasten the throatlash.
- Undo the quick-release knot, leaving the rope through the weak link.
- Undo and remove the headcollar.
- Pass your right arm through the left rein so you do not let the horse loose.
- Secure the headcollar through the tying-up ring – do not leave it hanging down on the ground as it then becomes a trip hazard.
- Bring the reins forward over the horse's head. Standing on the nearside (left) of the horse, by his shoulder, hold the reins in the right hand, 8cm (approx. 3in) behind the bit, with the index finger separating the reins.
- Hold the rest of the rein, and the whip, in the left hand.
- Keep the reins clear of the ground. The rein should never be wrapped around the hand, as this could cause serious injury should the horse pull away.
- Look straight ahead and lead and turn the horse as described in Section 5.

If the horse is wearing a martingale, the reins should only be taken forward over the head if the martingale is removed by undoing the rein buckle, releasing the martingale rings which can then be knotted at the chest. The reins would then be re-buckled. Alternatively the martingale can be left in place and the horse led from the reins which remain over the neck.

10.3 Prepared for leading the horse with a martingale

For further details of leading at walk and trot, and turning the led horse, refer to Section 5.

TACK-CHECKING PRIOR TO MOUNTING

Equipment notes:

You must:

Wear a correctly fastened crash cap.

Remove gloves for adjusting tack – put them in your pocket.

Put gloves back on prior to mounting.

Carry a whip which must not exceed 75cm (30 in).

The horse is most likely to be wearing a snaffle bridle with a cavesson or Flash noseband and a general-purpose saddle.

Based on what you learnt in Section 3, check that the tack is fitted correctly. Make adjustments as you think necessary. Make sure all straps are tucked into their keepers.

Checking the girth

The reins can now be put back over the horse's head in readiness for mounting. Check the girth by sliding your hand in between the girth and the horse's side in the direction of the horse's coat, i.e. from front to back. Never insert your hand against the lie of the coat as this may cause the horse discomfort. When tight enough, the girth should feel very 'snug' against the horse's side, such that your hand feels comfortably 'squashed'.

SAFETY TIP

- If you mount with a loose girth the saddle will slide around the horse's side and will need to be undone and re-fitted.

If your hand fits in easily without feeling 'snug', the girth is probably too loose. If you cannot squeeze your hand in at all, it is too tight.

At this point you should check that the numnah is fitted well up into the arch of the pommel of the saddle. If the saddle has slipped back or the numnah has slipped down too close to the horse's withers you should undo the girth completely, lift the saddle and numnah, pull the numnah forwards and refit correctly before fastening and tightening the girth.

Hook your arm through the rein to maintain control whilst leaving the hands free and gently tighten the girth as necessary before mounting. Avoid jerky movements which will cause discomfort and irritate the horse.

It is correct to use either the first and second or the first and third girth straps – never use the two rear straps together. The first girth strap is secured to the saddle independently of the second and third straps, acting as a safety feature should a retaining girth strap break.

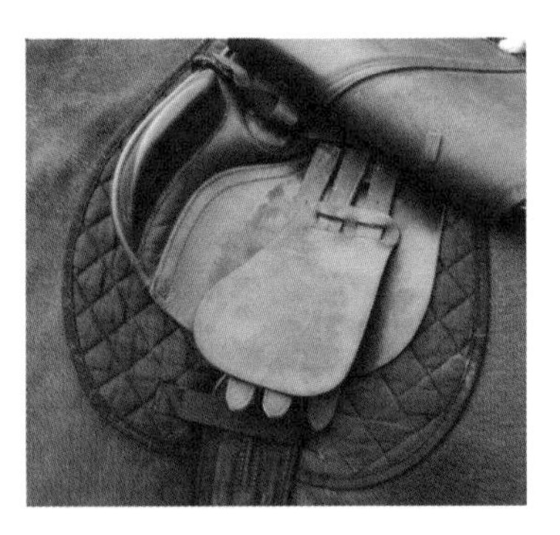

10.4 Correctly positioned buckle guard

Ideally, both sides of the girth should be done up equally and both buckles should be level to prevent uneven strain on the girth. Make sure the buckle guard is pulled down over the buckles.

Adjusting the stirrups

Once the girth is secure you should estimate and adjust the length of the stirrup leathers.

Pull the left stirrup down from its run-up position and, with your left arm hooked through the rein, estimate the correct length of leather by lifting the skirt of the saddle which covers the stirrup bar and placing your knuckles at this point. When adjusted, the base of the stirrup iron should reach into your armpit. This will give the approximate length of stirrup leather required. Make a note of the stirrup leather hole number, if numbered. Repeat this on the offside with the right stirrup, using the same hole number.

Take this opportunity to check that the stirrup bars are in the open position. The stirrup bars should never be closed when riding as the leather needs to be able to be pulled free in the event of the rider's foot becoming caught in the stirrup iron.

Once mounted you will make further adjustments to the length as necessary.

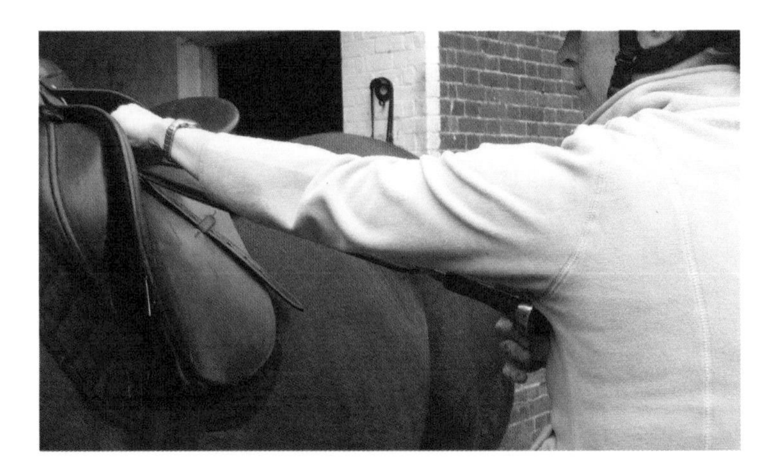

10.5 Estimation and adjustment of stirrup leather length

MOUNTING AND DISMOUNTING

Equipment notes:

Check tack, estimate stirrup leather length and tighten girth as necessary before mounting.

Wear a correctly fastened crash cap.

Put gloves on prior to mounting.

Carry your whip in your left hand.

Mounting

Mounting from a block is much easier for the rider, especially if the horse is tall and the rider less than athletic! Most importantly, it reduces strain on the horse's back, and on the tack.

Position the horse so you can mount from the mounting block on the horse's nearside. Turn to face towards the horse's tail.

1. Place the reins and whip in the left hand, just in front of the withers. The reins should be of equal length and short enough to prevent the horse from moving off, but not so short that they cause the horse to step backwards.

2. Grasp the left stirrup iron with the right hand and turn it out, or clockwise. Place the left foot, toe down, in the stirrup and turn to face the side of the horse.

3. Standing close to the horse's side, place the right hand on the waist of the saddle. (Holding onto the front or the back of the saddle when mounting can twist the saddle tree).

4. Spring up, lifting the right leg up and over the horse's back, taking care not to kick the horse's hip. Land lightly in the saddle. Once mounted, place your right foot into the right stirrup. To do this correctly, with the stirrup iron facing forwards, turn the front of the iron outwards, away from the horse and place your foot in. If you turn the iron the opposite way the stirrup leather will be twisted. It will dig into your leg and be counted against you during your exam if not corrected.

If mounting from the ground, more spring is needed and you need to have a longer left stirrup leather – e.g. if you have adjusted the leathers using the 'arm' measurement, you may find the left stirrup too high if trying to mount from the ground. It will need to be let down further and readjusted once you have mounted, but otherwise the procedure is the same.

ITQ 10.1

Why should a mounting block ideally always be used when mounting?

Tack checks once mounted

Checking the girth

Once mounted the first thing you should do is place your right foot in the right stirrup before checking and tightening the girth as necessary. Once mounted, the rider's weight can cause the girth to loosen.

Whilst preparing to mount you will have noted how evenly the girth is adjusted on both left and right sides. You should aim to have the girth adjusted as evenly as possible, ideally approximately halfway up the girth straps on either side. This will depend how well the girth fits the horse – if it doesn't fit very well, point this out so that an alternative can be provided.

To check the girth when mounted, lean forwards and slide your hand between the girth and the horse's side. This can be done on either side but if tightening the girth on the left-hand side, it makes sense to check the girth from the left-hand side too.

If it needs to be tightened on the left-hand side, proceed as follows:

- Keeping your foot in the stirrup, raise your left leg forwards, clear over the panel of the saddle.
- Hold the left saddle flap under the thumb of your right hand. Pull the buckle guard up out of the way of the buckles with your left hand.
- With your fingers pointing downwards, use your left hand to pull the end of one girth strap upwards. Use your index finger to guide the buckle pin into the hole. Repeat with the other girth strap, making sure they are evenly adjusted.
- Pull the buckle guard back down into position.
- Let go of the saddle flap and bring your leg back into the correct position.
- If in doubt, check the buckle guard is down on the right-hand side.

Adjusting the stirrups

You will have a general idea of your correct length of stirrup leathers for flatwork but, if you are riding a new horse for the first time, you may need to adjust your leathers as you get a feel for his shape and movement. If the leathers feel too long, too short, or uneven, you must adjust them – ideally, as soon as you move off, or otherwise as soon as you become aware of the need to do so.

At halt, take the main weight out of your stirrups but keep your feet in them when adjusting the leathers. Use the left hand to adjust the left leather; right hand to adjust the right one.

Once you have made the necessary adjustments, pull downwards on the back of the stirrup leather to ensure the stirrup leather buckle is right at the top, touching the stirrup bar. The free end of the leather must then be tucked back underneath your thigh and passed through the stirrup leather loop.

Ensure the stirrup leather is not twisted by turning the front of the iron outwards as you place your foot into it.

During the time spent checking the girth and adjusting the stirrups, the reins should be held in one hand with sufficient contact to dissuade the horse from moving off, but not so tight that he is irritated and provoked into fidgeting or stepping backwards.

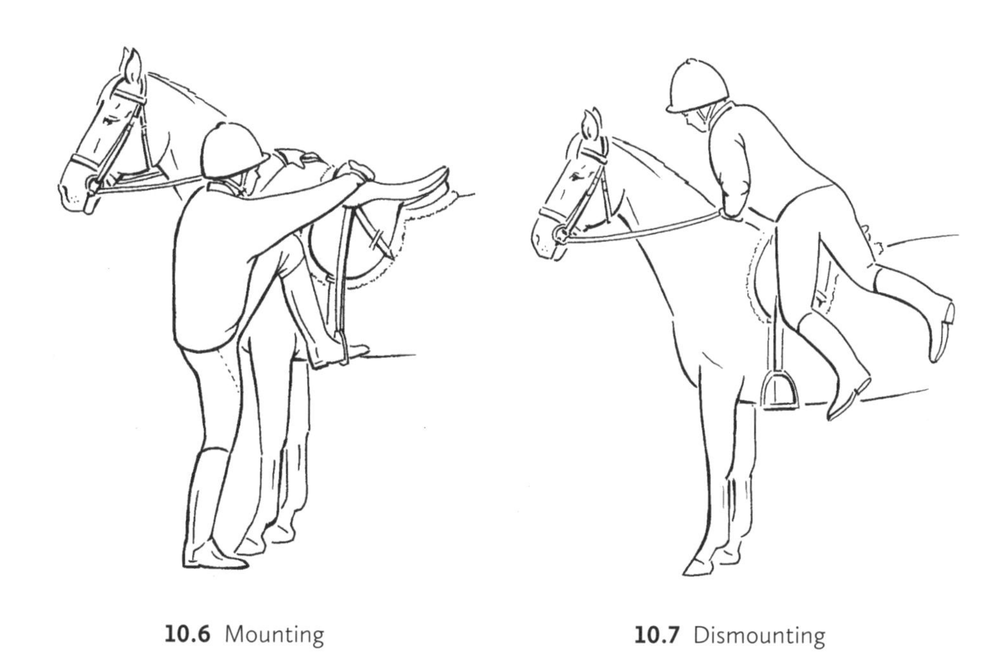

10.6 Mounting

10.7 Dismounting

Dismounting

Whilst it is considered correct to dismount on the horse's nearside (and this is what you will do in your exams), it is useful to practise dismounting from both sides.

To dismount from the nearside:

1. Take both feet out of the stirrups. The reins and whip are taken in the left hand close to the withers.

2. Place your right hand on the front arch of the saddle and lean slightly forward. Swing your right leg up and over the horse's back, clear of the cantle.

3. Drop down in a controlled manner, bending your knees as your feet touch the ground to help absorb the concussion of landing – a particular consideration when riding a big horse and/or dismounting onto a concrete yard.

To dismount from the offside, the procedure with arms and legs is reversed.

Running up the stirrups

- To prevent stirrups getting caught on gateposts or doors when returning the horse to his stable running up the stirrups is the first thing you must do as soon as you have dismounted.

- With the reins still over the horse's neck, hook your left arm through the left rein so your hands are free to run the left stirrup up first.

- Slide the stirrup back up the leather as far as it will go and feed the leather through the stirrup iron. Repeat on the offside.

- The girth can be loosened a couple of holes.

- Take the reins over the horse's head and position yourself in readiness to lead.

11 Seat and Balance

REQUIRED SKILLS/KNOWLEDGE	Learnt, revised, practised?	Confirmed
Maintain a secure and balanced position that is independent of the reins when riding with stirrups.		
• Walk, trot and canter in a secure and balanced position.	☐	☐
• Maintain a suitable rein contact.	☐	☐
• Use a suitable length of stirrup leather.	☐	☐
• Prepare for, and apply the aids to help ensure a correct strike-off into canter.	☐	☐
Maintain a secure and balanced position that is independent of the reins, when riding without stirrups.		
• Walk and trot in a secure and balanced position without stirrups.	☐	☐
Maintain a balanced, light seat in preparation for jumping and working over poles on the ground.		
• Maintain a balanced position at trot and canter in the light seat, with correctly adjusted stirrup leathers.	☐	☐
• Maintain a balanced position when trotting over poles on the ground.	☐	☐

THE CORRECT RIDING POSITION

Why correct position is important

The correct position is needed:

1. To balance the rider over the horse's own centre of balance, so there is no feeling of a need to grip in order to remain in the saddle at walk, trot and canter.

2. To harmonise the rider with the movement of the horse to avoid unintentional interference with the horse's action, i.e. the rider should 'go with' the horse.

3. To establish a level of communication acceptable to the horse, through the seat, legs and reins, enabling the rider to apply the aids correctly.

The correct position in the saddle is similar to the position that we take when standing. The body is held vertically, with the legs underneath and supporting the body. This natural and balanced position is also correct when riding a horse on the flat, with the minor alteration of a slight bend in the knees.

If an imaginary plumb-line were dropped from the rider's ear, the correct alignment of the body would be: ear in line with shoulder – in line with the elbow – in line with the hip – in line with the heel.

The joints of the ankles, knees and hips act as shock absorbers – the capacity to absorb shock is at its highest when the joint is slightly flexed. Setting or forced fixing of the position causes tension in the body which will prevent the rider from 'going with' the horse.

11.1 Correct riding position showing shoulder, hip, heel alignment

The head

You should look up and ahead in the direction of movement in order to maintain balance. Looking down, collapsing the chest and rounding the shoulders all affect balance and make your position less effective.

To help maintain correct head position, imagine:

– a full length mirror in front of you – look your reflection in the eye.

– a piece of string attached to the top of your crash cap, pulling you upwards.

Loss of balance is nearly always caused through tipping forwards, which results in the rider's weight being lifted off the seat bones and onto the front arch or fork only. To correct this, exaggerate looking up; this will raise and correct the body angle. The head and shoulders are the heaviest parts of the rider's body; any tilting of the head, either forwards or to one side will cause loss of balance.

Your head should not tip forwards, backwards or to the side. The chin should be raised so that the jawline is perpendicular to the ground. To help prevent your chin jutting forwards, think of bringing the back of your neck back towards your collar.

Riders often feel compelled to look down at the horse's neck – dropping the whole head alters weight distribution and affects your balance. As long as the head stays up, you can look down if necessary by moving your eyes only.

There should be no tension through the jaw, neck or shoulders, as this will transmit to other parts of the body as well as to the horse.

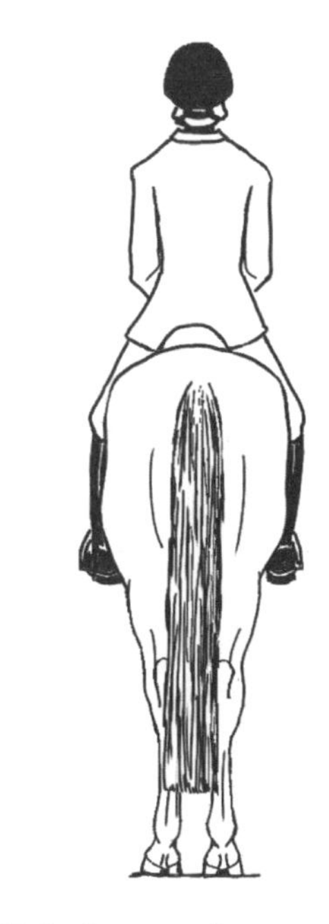

11.2 Correct alignment of the shoulders – square and level

The shoulders

The shoulders should be held square and level, with a slight feeling of drawing the shoulder-blades together, whilst avoiding tension. It is easy to become tight and stiff when trying to do this so it is often better to get the same result by thinking of raising and opening the chest.

Tightness in the shoulders affects the back and arms, which will impede the horse's free forward movement.

If you are naturally round-shouldered you will need to work on improving your posture, otherwise this rounding will cause tipping forwards in the saddle, which will spoil your position and weight the horse's forehand. A round-shouldered rider often also has a rounded back, which then cannot be used effectively in absorption of movement or as an aid.

The back

When riding, your back should be 'straight' yet supple. 'Straight' is in inverted commas because the spine naturally curves inwards at the small of the back. This doesn't mean to imply that the back should be hollowed or rounded, but should maintain its normal slight 'S' shape. When sitting in the correct position you will be able to follow the horse's movement, using the muscles in your back in a 'concertina' effect to absorb the horse's movement. An incorrect position limits the use of the back. When the spine is following its natural arch or curve, the rider's pelvis shows a slight forward rotation.

The seat

The seat provides the support for the body when the rider sits in the saddle. The rider effectively sits on a bony tripod or 'three-point seat', made more comfortable by the fleshy tissue-covering in that area (and hopefully a comfortable saddle!).

The 'three-point seat' consists of the two seat bones and the front arch (or crotch). Sit in the deepest part of the saddle, which is nearer the front than the back. Sitting on the back of the saddle will cause you to be 'left behind' as you will not be over the horse's centre of movement.

Aim for a feeling of closeness between the seat, thighs and inside legs with the saddle and the horse's sides. To help achieve this, open your hips as wide as possible to enable your legs to drop down naturally.

The aim is to develop an independent seat – a seat that can be maintained independently of the reins or any other false safety net. Gripping causes tension and compromises the seat.

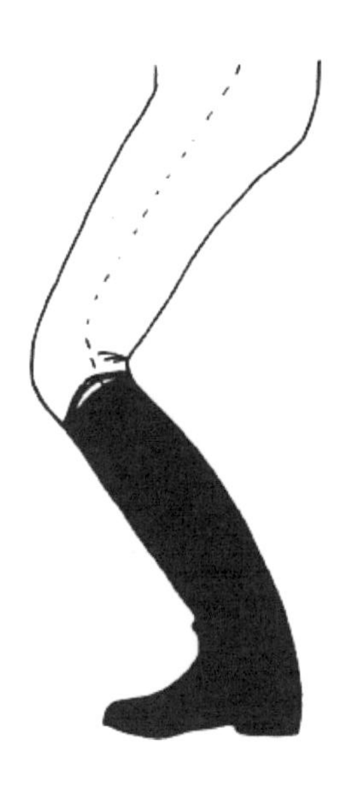

11.3 The rider's leg position

The legs

As mentioned, the legs should descend naturally from open hips. The hips play an essential role in the absorption of the horse's movement. Imagine your hips are twice as wide as normal so your legs can drop down around the horse's sides without undue stretching.

The legs should be directly under you, as if you were standing with slightly bent knees. This will ensure the alignment of the ear–shoulder–hip–heel. When not in use, the legs should stay lightly in contact with the horse's sides, not gripping but soft and relaxed.

Differences in rider conformation will affect the ability to take up the ideal position. For example, the short-legged rider with broad thighs will find it less easy to keep their legs close to the horse, but with practice a perfectly good position should be achievable.

With regard to function, the leg can be divided into two:

1. The thigh is an integral part of the seat, acting to hold the seat in position.
2. The lower leg applies the aids. It serves no purpose in gripping and will actually act to force the rider out of the saddle if used in this way.

Riders often grip up when first learning to maintain their balance and use their legs. To correct this tendency to grip, think of descending the knees as low as possible, as if kneeling.

The position of the lower leg can affect your overall balance. If it moves too far forward it will cause the upper body to become behind the movement; too far back and the upper body will tip forwards.

The foot is the lowest point of the riding position, with the stirrup holding the weight of your leg. Generally most of your weight is supported on the seat with just sufficient weight bearing down on the stirrup to keep it in place on the ball of the foot.

This weight through the leg travels down and into the heel as the lowest point. The deep heel position should not be forced as this will cause tension in the body. Instead, gravity should be allowed its influence. The fact that the heel is lower than the toe tightens the calf muscle at the back of the lower leg which, in turn, helps you give clearer leg aids, because of the tone of that muscle.

The arms and hands

The arms

The arms should hang naturally down by your sides, the elbows brushing against the outside of the pelvis.

The bend in the elbows permits a straight line to be drawn along the forearm, wrist, hand and rein to the bit in the horse's mouth. This is the rider's direct line of communication, which should not be broken by faults in the position of the arm, wrist or hand.

The hand should be positioned 'thumb on top' to allow flexibility of the elbow and ensure that the bones of the forearm (the radius and ulna) are parallel, with the radius above the ulna. Turning the hand over, as if to play the piano, results in twisting of the bones of the forearm and restriction of the use of the elbow.

Holding the reins

The reins are held in a gentle fist, the fingers closed lightly around the reins.

The reins enter the hand between the little and ring fingers, come over the palm and up between the thumb and index finger, where the excess is allowed to drape down over the horse's neck. The tightness of the fist will vary, but imagine that you are holding a small bird in each hand. You don't want the birds to fly away, yet you mustn't crush them. The fist can open slightly to allow the horse forward, or can close in a resisting action to prevent or check forward motion.

The hands must never come back towards your body, or pull, as this is a negative attitude and will result in the horse pulling against you. Human beings rely on their manual dexterity in almost every other walk of life, but this must be re-trained when learning to ride to give the seat, legs and back priority rather than relying on the hands.

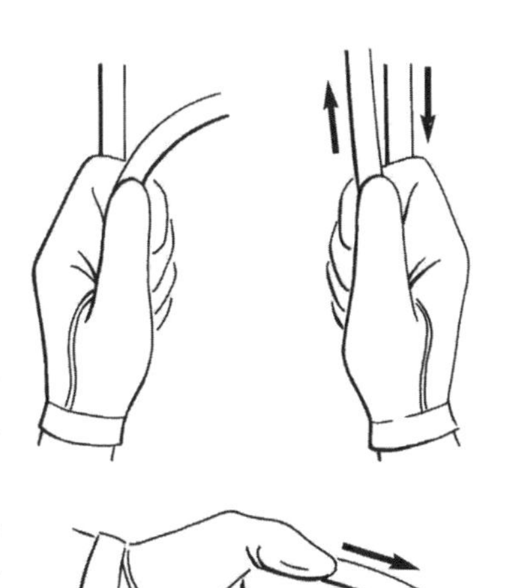

11.4 Holding the reins correctly

Functions of the hands

The hands assist the seat and legs to unite the horse and have two main functions:

1. **Passive resistance** – when slowing or stopping, the hands cease to 'allow' and reflect the blocking of further forward movement employed by the seat and back.

2. **Allowing** – the seat, legs and back ask the horse to move forward, and the hands permit the horse to obey by allowing – opening the fist slightly.

At all other times you must keep a steady, even yet 'elastic' (not fixed or rigid) contact with the horse's mouth.

Riding without stirrups

Riding without stirrups in the exam demonstrates your balance and security (or lack of it!). When instructed to ride without stirrups you must halt your horse in line, a safe distance away from other horses. Take your feet out of your stirrups, remove the leather from the stirrup leather loop and pull the buckle of the right-hand stirrup leather down approximately 10cm (4in). Cross the stirrup leather and iron over onto the left-hand side of the horse's neck/shoulder. Turn the leather so it lies flat and untwisted in front of your thigh.

Repeat with the left leather and iron, crossing this over onto the horse's right shoulder. Keeping the left iron on top means that if you have to dismount (or fall off) and need to remount, the left stirrup is easy to pull down.

When riding without stirrups it is important to maintain a balanced position, allowing your seat to absorb the movement and your legs to stretch down, without gripping, whilst keeping the toes lifted slightly. When trotting without stirrups you will maintain sitting trot.

PREPARATION FOR JUMPING

The rider's jumping position

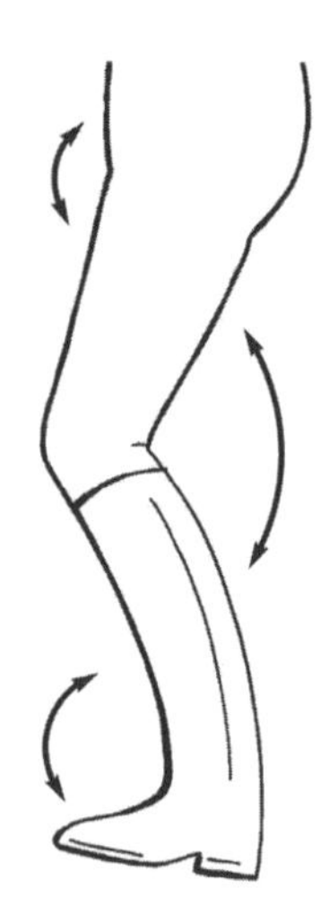

You will need to shorten your stirrup leathers by two or three holes from your flatwork length, to enable you to balance over the horse's forward-moving centre of motion and enable you to lift your weight off the saddle. Shortening the stirrup leathers closes the angles of the ankle, knee and hip joints.

If your leathers are too long you will have difficulty maintaining the jumping position, your lower leg will be insecure and you will probably lose your stirrups. If they are too short, you will be unable to use your legs properly to send the horse forwards and will find it more difficult to balance. Additionally your muscles will become tired very quickly.

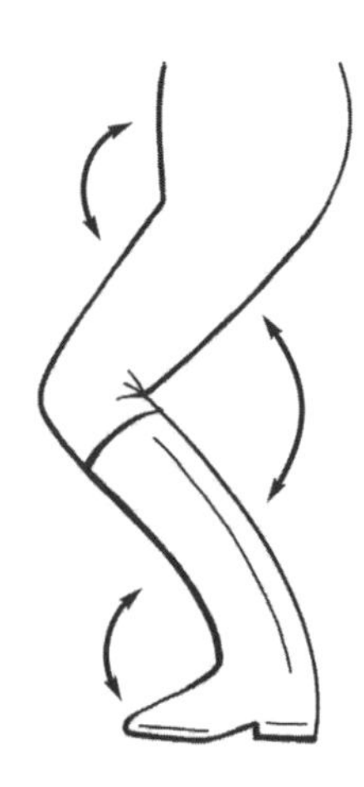

11.5 Closing the joint angles of the leg

The approach position

When adopting the approach jumping position the seat should move very slightly towards the back of the saddle as you fold forward from the waist, placing your body in front of the vertical plane at an angle of 30 or 40 degrees – the back should remain straight and the shoulders open (not rounded).

If you do not push the seat to the back of the saddle, but merely fold forwards, the horse's shoulders and neck will be beneath you, causing you to unbalance the horse's forehand and get 'in front of the movement'. This term describes what happens if the rider simply leans forwards, over the horse's neck, thereby being ahead of the horse's centre of balance. There will also be a tendency for the legs to move back, which causes a loss of balance, and further tipping forward.

For maximum security of position, your knees should come into contact with the knee rolls – the built-up area at the front of the saddle flaps, and your heels should be well down.

When in the approach position at walk, trot or canter you must look ahead and take the forward position. In trot, cease rising – the weight is taken on the thighs and stirrups and the seat 'hovers' lightly above the saddle. Keep your heels well down to maintain a secure lower leg position This approach position makes it easier to 'go with' the horse as he takes off.

11.6 The approach position

Pole work

You will be instructed to trot over poles on the ground, which are likely to be positioned down one long side of the school or on a long diagonal line.

Look towards the poles in good time, making sure the horse is straight, moving towards the centre of the line of poles. If the poles are set correctly, the horse should place each foot in the middle of the gap between the poles. Maintain the approach position over the poles. As you move away from the poles you can take rising trot again. Keep the horse straight and maintain a regular rhythm at all times.

Aids, Gaits and Riding in Company

REQUIRED SKILLS/KNOWLEDGE	Learnt, revised, practised?	Confirmed
Understand the natural aids.		
• Use natural aids for riding the horse forward on circles, turns and straight lines.	☐	☐
Ride on the correct diagonal.		
• Use the correct diagonal for rising trot.	☐	☐
Handle a whip that does not exceed 75cm (30in).		
• Handle a whip correctly when leading in hand, mounting and dismounting.	☐	☐
• Handle correctly a whip when adjusting reins, stirrups and girth.	☐	☐
• Handle or use a whip correctly when riding, changing the rein and working over poles on the ground.	☐	☐
Ride safely and in harmony with the horse and in company with others.		
• Build up a basic rapport when riding.	☐	☐
• Know the rules of the school when riding with others.	☐	☐
Know the footfalls of the horse's gaits and recognise when a horse is balanced.		
• Understand footfalls and beats of the horse's gaits.	☐	☐
• Recognise when a horse may be unbalanced.	☐	☐

THE AIDS

The aids are the means of communication between rider and horse and can be regarded as a language which the rider teaches and the horse learns. The aids are applied through varying pressures with the legs, seat and hands. The horse learns to yield or respond to this vocabulary of pressures. The reward for correct yielding or response is a cessation of the aid.

A trained horse will respond to refined aids which enable him to be light and responsive to ride.

The aids control and mediate:

- The rhythm.
- The energy, impulsion or desire to go forwards.
- Straightness.
- The bend.
- Turning.
- The gait – which includes halt and the transitions between and within gaits.

The aids can be divided up into natural and artificial:

The natural aids	**The artificial aids**
Seat	Whip
Legs	Spurs
Back	Martingales
Hands	Various gadgets
Voice	

In the early stages of learning to ride you will use natural aids and should only ride with artificial aids once balance and coordination are established and you understand the need for and use of such aids. However, a brief description of the main artificial aids is given later, by way of introduction.

Seat influences

The seat can help initiate, maintain and control impulsion. It also helps establish the horse's correct outline by assisting the engagement and drawing under of the hind legs with the subsequent rounding of the horse's back.

It also plays an essential role in determining direction of movement through the use of weight. When the rider is in the correct central position, the upright position of the pelvis gives the legs freedom to operate. Through turning of the pelvis (*not* tipping) the horse is encouraged to turn. This is described fully later on, see Turns and Circles.

Leg influences

The legs stimulate impulsion by activating the horse's hindquarters. If the hindquarters are inactive, the horse will seem to pull himself along from the shoulders, instead of the desired pushing forward from the hindquarters. To produce more activity in walk, the legs are used alternately, in time with the lateral steps of the gait; in trot and canter they are used together – although in canter the outside leg will be a little further back. The legs are also used to produce bend in the horse's body and indicate changes of direction.

The legs should give changing pressure when required against the horse's side, as constant pressure will be resented and later ignored. The leg aid acts by a closing in or nudging action, not a kick. The legs should not be moved further back on the horse's body if there is no initial response, as this puts you into the wrong position, which will confuse the horse and affect your balance. The inside of the leg, from under the knee to the instep of the foot, is used to give the aid. The back of the heel is not used, as this turns the knee and toe out, forcing the rider out of the saddle.

Back influences

As previously described, the back can be used to encourage the horse forwards or to slow him. It can quietly follow the movement of the horse through its concertina action, enabling the horse to move freely under the rider. It can be used more strongly, with the straightened back to drive the horse forward for extension or to encourage a horse who is thinking about refusing a jump (both uses requiring that the rider's seat is fully in the saddle).

Hand influences

The hands channel the impulsion produced in response to the leg and seat aids. They help regulate speed – the outside hand being chiefly responsible for this when riding in an arena. The hands also work as a pair to indicate direction – the inside hand asking for the bend, the outside hand supporting with the rein against the neck. Used in this way the reins are referred to as the **direct** and **indirect rein** respectively. It is useful to consider the hands as an extension of the seat.

- The reins should be even (i.e. same length), held with thumbs uppermost, maintaining a light and even contact.
- There should be an unbroken straight line running from the bent elbow, along the forearm, wrist, hand and rein to the bit.
- The reins must *never* be used to aid the rider's balance.
- The horse should be ridden forwards into the contact; the contact should not be taken back to the horse.

The voice

The voice is a very useful aid – it can be used to calm, reward and correct and is important in the initial training of the young horse. It is the *tone* of voice used, rather than the actual words which the horse will listen to, although some riding school horses seem to recognise words used by the instructor, which can leave the rider as an unwilling passenger. For example the instructor may say 'Prepare to walk' and before the rider can respond the horse is walking.

The voice should not be used as punishment; shouting at a horse is likely to upset him rather than correct him. A raised voice tends to indicate fear in the rider or handler.

Artificial aids

The whip

A whip is carried to back up or reinforce the leg aid. It is not used to punish the horse.

There are two types of whip used for general equitation:

Schooling whip – usually about 1m (39in) long, used only for flatwork (not used in Stage 1).

Jumping whip – 75cm (approx. 30in) long, used when showjumping or cross-country jumping.

Novice riders will learn to carry a short whip before using a schooling whip. In the Stage 1 exam your whip must not exceed 75cm (30in).

When first learning to ride you will not carry a whip as this presents too many problems when learning how to hold and use the reins, the effect of the legs and basically trying to stay on the horse. Once you are fully balanced and can maintain a steady contact with the horse's mouth, you should learn how to carry a whip.

When working in the school the whip is carried in the inside hand, resting across your thigh, to support your inside leg and help discourage the horse from 'falling in'. Only a short length of whip handle should protrude at the top of the hand to avoid injury should the horse throw his head up suddenly.

Each time the rein is changed the whip should be changed to the new inside hand. The method of changing a short whip (which is what you will probably start with) from one hand to the other is:

- Put both reins in the whip hand (the hand holding the whip).
- Take hold of the top of the whip with the spare hand and pull the whip through.
- Retake the reins in both hands and position the whip so it rests across your thigh.

When needed, the whip is used once immediately behind the rider's leg. The whip should not be used further back than this as it will not be reinforcing the leg aid and will probably cause the horse to buck or kick out.

To use a short whip, the reins should be taken into the outside hand and the free hand used to apply a light tap with the whip. This prevents the horse from getting jabbed in the mouth. Allow the horse to go forwards once the whip has been used.

When adjusting the girth or stirrup leathers, the whip is held in the hand not being used for this purpose, together with both reins.

Spurs

Only experienced riders with a secure leg position should wear spurs. They should be regarded as a refinement of the leg aids and should never be worn pointing up into the horse's sides. You will not wear spurs in your Stage 1 or 2 exams.

Running martingale

A running martingale can usefully be worn by the novice rider's horse. Its main function is to prevent the horse carrying his head too high by exerting pressure on the bit when the horse raises his head beyond the point of control. The neckstrap can be held by a beginner rider to aid balance. For beginner riders, a correctly fitted running martingale can give a degree of regulation to the rein aids, but emphasis must be placed on establishing balance and still hands before the pupil rides independently holding both reins.

'Gadgets'

Balancing reins, draw reins and other such gadgets have no place in the tack worn by a horse ridden by beginners or novices. It is a point of contention with some trainers as to whether they have a place at all and it is beyond the scope of this book to debate the subject – suffice to say that in the hands of an inexperienced rider such gadgets can do irreparable harm to the horse, both physically and to his mental attitude to work.

'On the aids'

Whilst we are discussing the aiding system it is relevant to look at what this term means.

A horse is said to be on the aids when he appears alert, yet calm and trusting – ready and able to carry out the rider's next request with ease and enthusiasm. Not every horse presented in the Stage 1 exam will be completely on the aids; you must practise riding a range of horses, including slightly staid, even 'lazy' horses, so you are effective on these types.

ITQ 12.1 ?

List the natural aids and explain how they are used for communication between the rider and the horse.

TRANSITIONS, STRAIGHT LINES AND TURNS

Preparing for transitions

To increase the chances of making a smooth and clear transition you must first ensure that the horse is moving actively forwards in an even rhythm.

Before making any transition you should give the horse a preparatory signal. As you progress in your training this is often referred to as a 'half-halt'; it helps balance and prepare the horse for another instruction. Very simply, before you want to make your actual transition you need to get your horse's attention and make sure he is prepared for, and therefore 'listening' to the upcoming aids. To do this you should make sure you have sufficient impulsion by using your legs and seat, and as you do so, apply slight pressure to the reins (which is almost immediately released). This acts as if to say, 'Listen to me – I'm about to ask you to do something else…' Then follow on with the aids to make your transition.

You must prepare the horse slightly in advance of the arena marker at which you wish to make your transition. If you have been instructed to make a transition in between two markers, you should aim to do so in the middle of the two. Do not leave it too late to prepare or you are likely to miss the arena marker.

Upward transitions

Examples of upward transitions are: halt to walk, walk to trot and trot to canter. At Stage 1 level, all upward transitions are made **progressively** – that is to say, for example, that in going from walk to canter you will do so via a few strides of trot. **Direct transitions**, miss out the intermediate gait – for example, from walk directly into canter. These will be introduced at a later stage.

A good upward transition is active, smooth and positive. The horse must be attentive, ready to move immediately in response to your leg aids. Good preparation greatly increases the chance of making a clean transition.

To make a transition from halt to walk, maintain an upright position and squeeze with both legs simultaneously. The rein contact must allow the horse to move forwards. As he does so, the leg aids can cease, unless he is reluctant or inactive, in which case they should be repeated.

The aids for upward transitions into trot and canter are discussed under Riding at Trot and Riding at Canter later in this section.

Downward transitions

Examples of downwards transitions are: from canter to trot; trot to walk; walk to halt.

Until you are quite advanced, transitions from trot to halt or canter to walk are made progressively, i.e. through the intermediate gait. As with upward transitions, at Stage 1 you are required to make progressive transitions in the exam.

Direct downward transitions – e.g. canter to walk (missing out trot), or trot to halt (missing out walk) are more difficult to ride well than direct upward transitions and are not required at Stages 1 or 2.

To make a downward transition, apply a preparatory aid, then, keeping your legs close to the horse's sides to maintain impulsion, sit tall yet deeply into the saddle and brace your back slightly while applying light resistance with the reins. As the horse obeys your instructions release (but do not lose) the rein contact. You must practise riding downward transitions using the seat and back, making sure you never rely solely on the reins.

Keeping the horse straight

When moving on straight lines, e.g. when riding down the straight sides of the school, across the school on a diagonal, or up a centre or quarter line, the horse's body should be straight, without lateral bend in the spine and, on each side of his body, the hind foot should follow the track made by the forefoot. His head and neck should be straight, i.e. there must be no flexion in the neck and his head should not tilt one way or the other.

Straightness is achieved by maintaining a good level of impulsion and an even rein and leg contact. If the horse deviates from straightness, you should make the necessary correction using leg and/or rein aids. If the horse's quarters swing inwards or outwards, draw your relevant leg back slightly behind the girth and apply it to correct the deviation of the quarters. If the horse's neck flexes inwards or outwards, use the appropriate rein to make a correction. This will usually entail taking a slightly firmer contact with the rein on the opposite side to which the horses is flexing his neck. However, be careful that you do not *cause* the flexion by taking too strong a contact on the side to which the horse flexes.

The most common cause of lack of straightness is loss of impulsion and forward momentum – just as when riding a bicycle, it is easier to stay on a straight line when moving positively forwards; much more difficult when moving slowly.

Turns and circles

Turns and circles will be dealt with together because their aids are the same, i.e. the aids to ride a circle to the right are the same as to ride a turn to the right, and the principles for correct circling are the same as for correct turning.

The horse changes his direction of motion by turning either to the left or the right – his body following the shape of the turn laterally. For example, when the horse turns to the right, his right side contracts as it follows the inside of the bend and his outside lengthens to take the longer route through the turn.

The horse's body effectively takes on a slight curve from poll to tail when viewed from above. The horse's hind feet follow in the tracks of the forefeet and he looks in the direction in which he is travelling. The same is true when working on a circle.

The aids to turn and circle

In this example, we will assume the intention is to turn or circle right. To do so to the left, the aids are reversed.

As the horse is moving forward in the chosen gait, ask him to flex and look slightly

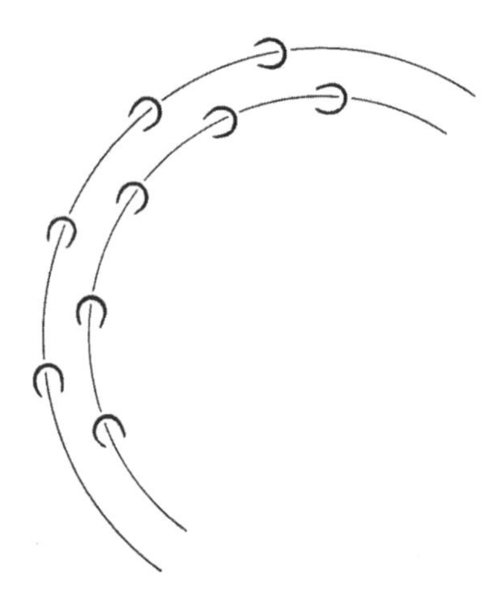

12.1 Working on two tracks

to the right by closing the fist of your right hand so you can just see the horse's right eye. The outside (left) rein limits the bend and speed, and controls the horse's outside shoulder. 'Controlling the outside shoulder' can be a difficult concept to understand when learning.

As the horse moves on a straight line or turns or moves on a circle he must travel on **two tracks**. This means that his right hooves are on one track, his left hooves on another. In the event that the horse's neck is bent too much to the inside, it can cause the horse's outside shoulder to drift outwards, termed 'falling out' or 'losing the outside shoulder'. The outside foreleg then takes up its own track, i.e. the horse moves on three tracks, which is incorrect in these situations.

To prevent this, a contact must be maintained on the outside rein; this stops the neck from bending too much to the inside and therefore controls the shoulder. The rider's outside leg behind the girth stops the quarters from swinging out. If the quarters swing out the outside hind leg takes up a third track and the horse does not genuinely bend through the turn or circle.

The rider's legs are used to 'bend' the horse by maintaining the inside leg (inside the bend of the body – in this case the right leg) on the girth, but pressing a little more weight down onto the right stirrup and seat bone, and drawing the outside leg (left leg here) back a little behind the girth to control the swing of the horse's quarters.

The inside leg on the girth helps to produce the bend but also acts to prevent the horse from falling in. The whip is carried in the inside hand to reinforce the leg aid if required.

Forward momentum must always be maintained and ideally there should be no loss of rhythm through the turn. If the rider's legs are not used correctly through a turn or on a circle, and the reins are relied on, the horse will be unbalanced, probably falling out through his outside shoulder or swinging his quarters out and the rhythm will be lost. The inside hand should be opened a little, but should not pull back, as this will constrict the horse's neck and 'block' the forward movement of his body, thus impairing rhythm and increasing the chance of the horse's quarters drifting out.

The smaller the circle or more acute the turn, the greater the degree of bend required from the horse. Therefore, in early lessons the novice rider is taught to ride 20m and 15m circles, and turns must be made simply through wide angles.

ITQ 12.2

What are the functions of the following when riding a turn or circle?

a. Inside leg:

b. Outside leg:

c. Inside rein:

d. Outside rein:

ITQ 12.3

What is meant by the following terms?

a. 'Working on two tracks':

b. 'On the aids':

THE GAITS

Walk

Walk is a four-time gait with no moment of suspension. The sequence of footfall at walk is inside hind, inside fore, outside hind, outside fore. The steps should be regular – equal in distance and time.

Walk is the easiest gait in which to practise and show correct posture. Avoid obviously 'pushing' the horse along with your seat; use your legs alternately to maintain impulsion, ensuring that your hands move gently in harmony with the horse's head movement in order to retain the correct contact.

Trot

Sequence and rhythm

The horse trots in two-time. If you listen to a horse trotting on the road you will hear a distinct 1-2, 1-2, 1-2 beat. The horse moves his legs in diagonal pairs, separated by a moment of suspension when all four legs are off the ground. The legs move as follows:

1	Left hind leg and right foreleg	**or**	Right hind leg and left foreleg
	Suspension		Suspension
2	Right hind leg and left foreleg		Left hind leg and right foreleg

The trot should be regular and unhurried, with the knees and hocks bending and the feet lifted cleanly off the ground in their diagonal pairs. The trot should be generated from the hind legs, sending the horse forwards, rather than the shoulders dragging the horse along.

12.2 The trot

The aids for trot

The horse naturally carries his head slightly higher than at walk when trotting, so you should prepare for this by shortening the reins whilst closing the legs around the horse's sides, to ensure that the horse goes forward into the bridle. Apply a preparatory aid, keep looking ahead and close your lower legs to send the horse into trot. Once the horse is trotting, the leg aids can cease as a reward but may need to be reapplied as soon as you sense any drop in activity – do not wait until horse has almost fallen into walk before giving him another leg aid.

Rising trot

As the horse moves into trot, remain sitting for the first three or so strides, before starting to rise.

Rising trot should feel as if you are allowing the horse's movement to lift you slightly out of the saddle, lifting your seat bones above the saddle on one beat out of two. You should not rise too high, or you will sit down too late and be behind the movement. If you are completely out of the saddle when rising, the horse will move forward underneath you and you may find yourself sitting too far towards the back of the saddle.

Visualise it as if going from sitting to kneeling and back again. When in the rising phase your weight is transferred to the thighs, knees and stirrups. To make sure that you stay with the movement try to feel that your rising is 'led' by your belt buckle, so that you bring your seat forward and down into the correct part of the saddle. Your legs shouldn't move backwards or forwards when rising. If the horse needs to be sent forward in trot, apply your legs by nudging in during the sitting phase of the rising trot.

Your hands should remain steady – there is no head-nodding motion in the trot of a sound horse. When learning it is a good idea to hold a neckstrap.

When riding in the school, you should sit as the outside diagonal comes to the ground. The diagonals are identified as inside or outside according to the foreleg, e.g. on the right rein the outside diagonal is the left foreleg and right hind leg.

When on the correct diagonal you should rise as the horse's outside shoulder moves forward away from you and sit when the outside shoulder comes back towards you. To begin with you can glance down at the shoulder but you should eventually be able to tell by feel whether or not you are on the correct diagonal.

If you are incorrect, i.e. rising when the horse's outside shoulder comes back, correct by sitting for one extra beat and then rising again. When the rein is changed, so is the diagonal to ensure that the muscles on both sides of the horse's body are worked equally. At each changing of the rein across the diagonal you should sit for an extra beat as you pass over X.

The reason for selecting the outside diagonal is to ensure that your seat is in the saddle when the horse's inside hind leg is on the ground, prior to propelling the horse forward. You can then use your back, seat and legs more efficiently to promote balance and impulsion. When hacking, the diagonals should be changed occasionally to avoid either horse or rider favouring one or the other.

Sitting trot

You will need to sit to the trot before a transition to walk or canter. For sitting trot cease rising and allow your weight to follow the movement of the horse. Remain upright, with still hands. Avoid gripping with your legs as this will cause you to bounce in the saddle.

EXAM TIPS

Never trot on the incorrect diagonal in your exam. Be aware of which diagonal you are on at all times and, if you are on the incorrect one, change it.

Remember to change diagonals when you change the rein.

Canter

Sequence and rhythm

The horse canters in three-time, i.e. you will hear and feel a 1-2-3, 1-2-3, 1-2-3 beat. There is a moment of suspension when all four feet are off the ground. The 'leading leg' is the last foreleg to come to the ground in the three-time sequence. On a circle, the horse's body bends in the direction of the leading leg. The sequence of footfalls for the right-lead canter is:

1. Left hind leg
2. Right hind leg and left foreleg together
3. Right foreleg

Suspension

The sequence always starts with the outside hind leg, which pushes the horse off into canter. The canter should be regular and unhurried, with active hocks and good engagement of the hind legs, i.e. the hocks should bend and come in under the horse's body to provide effective propulsion. The horse should be straight, with his forehand directly in front of his hindquarters. A common fault is for the horse to throw his quarters in to the inside, thus making him crooked.

The horse's head moves in time with the leading leg. There is a rocking motion to the canter, whereby the horse's quarters rise as the leading leg comes to the ground and then sink again as the head and neck rise. A horse who appears to keep his quarters in the air all the time in the canter is on his forehand, i.e. carrying too much of his

12.3 The canter

The aids for canter

During the transition to canter, the horse should look very slightly to the inside – being the side of the desired leading leg. This is most easily achieved if the canter strike-off is asked for in a corner of the school when the horse should be naturally bent to the inside.

The corner also helps to balance the horse for the transition to canter, making it more likely that he will strike off on the correct leg. As you become more practised you can progress to cantering on a named leg on a straight line without the assistance of a corner but for the purposes of your Stage 1 exam you must ask for the transition in a corner.

Plan which corner you are going to use and prepare several strides in advance. The rein length used for trot should not need altering but you should ensure that the trot is active and the horse attentive.

- Take sitting trot approximately 4–6 strides before your planned transition.
- Sit with the inside hip slightly in advance of the outside hip and allow the inside (inside the bend) leg to come deep as it stays at the girth. The outside leg (on the outside of the bend) is drawn back behind the girth.
- To make sure the horse is listening, give a preparatory aid just before your planned transition.
- Stay upright and look ahead during the transition – do not tip forward over the horse's shoulders. Apply both legs in this position and slightly open the fingers to permit the horse to obey. The contact is maintained as the fingers allow for the rocking motion of the horse's body. As in sitting trot, you must remain deep in the saddle, using muscle tone to stay erect and balanced whilst absorbing the motion through the concertina action of the back and joints. The legs can be reapplied to maintain the canter.

You will feel and follow sinking, advancing and lifting movements as the horse canters.

Checking the leading leg

At Stage 1 you must be able to identify the correct lead. If the horse is on the correct leg, the inside foreleg should feel as if it is advancing further forward than the outside foreleg. It is important to practise this. When learning it is acceptable, initially, to look for the canter lead. Do so by looking down towards the front of the horse's inside shoulder *without* leaning or tipping your head forwards. You should see the horse's inside foreleg stretching out further forward than the reach of the outside foreleg.

When you know you are on the correct leg spend time developing a feel for this. You are aiming to be able to tell which leg you are on purely through feel. As your training progresses you will ultimately be able to tell which foreleg the horse is going to lead with simply by feeling which hind leg has taken the first step of the sequence during the transition and, if it is the inside hind, you will be able to correct it before the horse has even taken one complete stride!

If the horse strikes off on the incorrect lead you must make a transition back to trot immediately (this proves to the examiners you knew you were on the incorrect leg), and prepare and ask again on the next corner. If the horse strikes off on the incorrect lead yet again, trot and bring him onto a 20m circle to ask again.

EXAM TIP

Never canter on the incorrect lead in your exam. If the horse is problematical you must keep trying to correct the canter lead.

ITQ 12.4

Explain the reasons for sitting on the outside diagonal when trotting.

ITQ 12.5

What are the aids for canter?

The unbalanced horse

The most common signs horses may show when they are unbalanced include any combination of the following:

- Inactivity of the hindquarters.
- Carrying too much weight over the forehand ('on the forehand').
- Falling in on corners or circles.
- Rushing.
- Breaking from the required gait, usually downwards.
- Holding the head too high or too low.

Common causes of loss/lack of balance include:

- Incorrect or insufficient training.
- Inappropriate riding technique (the rider's actions unbalance the horse).
- Bad ground conditions.
- Physical problems, including lameness.
- Incorrect shoeing/ need for re-shoeing.

RIDING IN COMPANY

Preparation and mounting

- As mentioned in Section 3, tack and equipment must be safe and well fitting. Accidents to both horse and rider can be caused by unsafe tack. Check for wearing, cracking and loose stitching.
- Always check the girth before attempting to mount. If it is loose the saddle will slip round as you put your weight into the stirrup iron.
- Always use a mounting block to reduce strain on the horse's back and prevent the saddle from being pulled and twisted. The block should be solid and well-built. Don't use anything flimsy such as a milk crate, which either horse or rider could put a foot through, or which may tip over.
- If mounting in a yard, the gate should be closed. If several riders are mounting care must be taken that they don't get too close to each other as there is a risk of the horses kicking. If there is a mounting block in or near the school, lead the horses to the school.

- Keep horses out of kicking reach of each other, particularly if one is known to kick. Known kickers should be placed as rear file and the ride warned of the horse's anti-social habit. A red ribbon tied into the top of the tail acts as a warning that the horse is prone to kicking. You must be alert to the warning signs of potential problems between horses, such as ears laid back and tail-swishing.

Riding in a class

As a ride

When behind another horse, a gap equal to at least half a horse's length (preferably a whole length) should be maintained between horses. When looking through the horse's ears, you should be able to see the tail of the horse in front. This will help prevent a kicking incident.

If your horse keeps catching up with the horse in front you must check your speed – are you pushing your horse on too quickly? If so, steady the horse to avoid catching up. If you are happy that you are maintaining the correct rhythm and speed, try to work deeper into the corners, to give yourself more space. If this is not enough you will need to circle inwards from your place in the ride to take up a safer position at the rear of the ride.

Careless riding and a lack of distance safety awareness will cause you to fail your exam.

If, in normal lesson situations, once warmed up, you wish to remove your jacket or jumper, you should ask your instructor and turn in off the track. If the horse isn't known to you or the instructor it is safer for you to dismount to remove your jacket in case the horse becomes upset. If you need to remove your crash cap to take off the jumper, you should certainly dismount. If the horse is known to be very safe and not to mind clothing being removed you should halt the horse and the instructor should then hold him. If the horse is not held, you are in a very vulnerable position when halfway through the removal e.g. when the jumper is over your head, or your arm is behind your back as you remove the jacket. If the horse takes fright at seeing clothes flapping out of the corner of his eye, a dangerous situation could arise.

SAFETY TIP

- When riding in closed order, you must never get closer than half a horse's length to the horse in front of you – preferably keep a whole length between horses.

In open order

Open order is the term used to describe a number of riders riding individually rather than together as 'a ride'. The main safety rules include:

- Always look up and ahead, i.e. where you are going, to avoid collisions.

- Plan your movements and be observant as to where other riders are going.
- Pass oncoming riders left-hand to left-hand and always allow enough room to pass without danger of the horses kicking. If you are on the right rein you must come far enough onto an inside track so that you allow the oncoming rider enough room between your horse and the wall of the school.
- Riders working in the slower gaits should take the inside track (i.e. come off the main outside track and work 2–3m inside it) thus leaving the outside track free for those working in canter, for example. A rider who requires the outside track can call 'Track please' so that the other riders can get out of the way.
- Always assume that all horses kick. Never get so close that horses could reach each other (or the rider) if they kicked out.
- No rider should jeopardise the safety of another by boxing them in between another horse and the wall, or between two horses.

ANSWERS TO IN-TEXT QUESTIONS

1.1 It is unsafe to leave a nylon headcollar on a horse who is loose in the stable because it may get caught on the top bolt of the door as the horse looks out. Nylon headcollars do not break and the horse would panic and be difficult to free.

1.2 The safest way to pick up a 20kg bag of horse food is to stand close to it and, keeping your spine straight and knees bent, grasp the sack securely and lift by straightening your knees rather than using arm strength alone. Use a sack barrow if moving it any distance.

1.3 To avoid back injury when moving bales of hay seek assistance and each hold one string to carry it. Keep your back straight. Use a barrow to move a bale any distance.

1.4 The safest way to carry water buckets is to have a bucket in each hand, each containing the same amount of water to keep the load balanced.

1.5 A bale of plastic wrapped shavings can be lifted in the same way as the feed sack but they do tend to be more awkward because of their size. Seek help and use a sack barrow.

1.6 It is important to speak when approaching a horse as he may not have seen you. Your voice alerts him to the fact that you are there – so you are less likely to frighten him. Always speak before you touch a horse, especially if he is not facing you.

1.7 Four safety points to observe when tying horses up:
1. Always use a quick-release knot.
2. Always use a weak link, attached to a secure object, as this will break in the event of the horse pulling back and panicking. If there is not a weak link the horse may injure himself.
3. Never leave a tied horse unattended.
4. Never tie horses up close together.

1.8 Two examples of unsafe tying practices:
1. Tying the horse to a fence rail. The horse could pull back and bring the rail with him.
2. Tying up with a long rope over which the horse's leg could get entangled.

1.9 Two points to remember when leading a horse through a doorway:
1. Make sure the door is fully open and cannot swing onto the horse.
2. Lead the horse straight through the door – not at an angle, when he would be likely to bang his hips on the door post.

1.10 When grooming the following brushes are used for:
a. Dandy brush – to remove dry mud and dirt from the coat.
b. Body brush – removes dust and grease.
c. Metal curry comb – cleans the body brush – not used on the horse's coat.
d. Plastic curry comb – removes dry mud, dirt and loose hair.

1.11 If the hooves are not picked out twice daily the horse may develop an ailment called thrush in the clefts of the frog. (Regular inspection of the foot will also alert you to other issues, such as a loose shoe or embedded stone.)

1.12 Quartering is a short grooming given before exercise to tidy the horse up.
Strapping is the full grooming given after exercise to tone the muscles and thoroughly clean the coat and skin.

1.13 Strapping is carried out after exercise because the pores of the skin are open, allowing oil in the skin to give the coat a shine. The dirt in the coat will loosen off more easily too.

2.1 You must never put a wet tail bandage on as it may shrink and interfere with circulation in the dock.

2.2 Tail bandages are applied:
- To protect the dock when travelling.
- To improve the appearance of a pulled tail.
- During covering or foaling.

2.3 The main disadvantage of using rollers and surcingles to hold rugs in position is that they exert continual pressure on the spine. This can lead to a sore back – a thick sponge pad should be used beneath both rollers and surcingles.

2.4 A correctly fitted rug will:
- Not be too tight at the shoulders.
- Extend to the top of the tail.
- Extend to just below the level of the elbows, covering the horse's abdomen.

2.5 The horse should not be left standing with only the belly straps of the rug fastened as there is a risk that the rug will slip backwards and become entangled around the horse's hind legs.

2.6 The horse should not be left standing with only the front buckles fastened as there is a risk that the rug could slip and become entangled around the horse's forelegs.

3.1 Refer to Figure 3.1 on page 34.

3.2 To tell if a cavesson noseband is at the correct height, you should be able to fit two fingers between the noseband and the projecting cheekbone. You should be able to fit two fingers between the horse's face and the noseband if it is tightened correctly.

3.3 The main purpose of the Flash noseband is to help stop the horse from crossing his jaw and opening his mouth.

3.4 The bit must not protrude more than 1cm (approx. ½in) either side of the mouth and must be high enough to create approximately two wrinkles – but no more – in the corners of the lips.

3.5 All saddles need to be re-flocked at least once a year as the flocking packs down and shifts, causing uneven lumps which create pressure points. Uneven pressure will give the horse a sore back.

3.6 You should fit a saddle without a numnah because the numnah will disguise the true fit of the saddle. All saddles should fit the horse without a numnah, although a numnah should be used when riding.

3.7 Four points to observe about a well-fitting saddle:
1. The saddle should sit level on the horse's back – it should not tilt forwards or backwards as this would make it difficult for a rider to maintain position and would exert uneven pressure on the horse's back.
2. The saddle should rest evenly on the lumbar muscles, which cover the tops of the ribs. It must not touch the loins. The full surface of the panels must be in contact with the horse's back to distribute the weight over the largest possible area.
3. The saddle must not pinch the shoulders. The knee roll, panels and saddle flap should not extend out over the shoulder, as this would restrict the horse's freedom of movement.
4. With the rider mounted there should be sufficient clearance beneath the pommel when in an upright, flatwork position and a forward, jumping position. You should be able to see daylight through the gullet. At no time should the saddle touch the horse's spine. When viewed from behind, the saddle should not appear crooked or twisted.

3.8 You should always try to avoid putting a saddle down on the ground – especially a new one! If you need to put a saddle down on the ground you must make sure that the leather does not get scratched. Lay the numnah on the floor and rest the saddle on its front arch against a wall on top of the numnah. Place the girth between the cantle and the wall. At no time should any leather part of the saddle touch the concrete yard or the wall!

3.9 The purposes of using a numnah are to absorb sweat from the horse's coat, which reduces the likelihood of friction, and to absorb and even out pressure from the rider's weight. Numnahs also keep the saddle clean, which reduces time spent cleaning tack.

3.10 A running martingale is used to prevent the horse from carrying his head high, beyond the point of control.

3.11 To tell if a martingale is fitted correctly:
- When the horse stands with his head in the normal position the martingale straps must be slack.
- It should come into action only when the horse raises his head beyond the point of control.
- Before passing the reins through the martingale rings, take both straps and hold the rings together towards the horse's withers – the rings should not touch the withers as this would make the fit too loose.
- Then pass the reins through the rings and re-buckle; standing beside the horse, hold the reins up in the position they would be in if the rider were holding them on the horse.
- Check the fit – the straps should be slightly loose.
- The neckstrap should be adjusted to admit the width of your hand.
- A rubber stopper must be used on the neckstrap to prevent it slipping along the main strap, which will affect its action. Rein stops are used on each rein to prevent the martingale rings running up too near the bit rings, which could panic a horse.

3.12 Three reasons for cleaning tack regularly are:
1. Cleaning keeps the leather supple and therefore less likely to crack.
2. Cleaning the tack gives you the opportunity to check it for safety.
3. Clean, supple leather looks and feels good to work with.

3.13 Three areas which need to be checked for safety when cleaning tack:
1. All stitching on the bridle.
2. Girth straps – check stitching and leather for signs of stretching and cracking.
3. Stirrup leather – check stitching and leather.

3.14 Two causes of saddle sores:
1. Ill-fitting saddle – too low on the withers, or lumpy stuffing.
2. Dirty numnah and/or horse.

3.15 If a horse has a saddle sore, the cause must be removed. Don't continue to work the horse in ill-fitting or dirty equipment. Clean the area, using salt water, and apply a poultice. Use kaolin if the skin is not broken, Animalintex if the skin is broken. Rest his back until it has healed properly.

4.1 Five reasons why bedding is necessary:
1. Encourages the horse to stale.
2. Provides warmth and reduces draughts.
3. Prevents jarring of the limbs.
4. Encourages the horse to lie down and rest.
5. Prevents the horse from slipping on the floor.

4.2 Two disadvantages of using rubber matting as bedding:
1. Rugs become badly soiled.
2. Some horses are not happy to stale directly onto rubber matting.

4.3 Bedding suitable for an allergic horse:

1. Rubber matting.
2. Shredded paper.

4.4 Straw is not suitable for an allergic horse because of its dust content.

4.5 Two advantages of the deep litter system:
1. Saves labour on a daily basis.
2. Saves on bedding materials as less clean bedding is needed each day.

Two disadvantages:
1. Every few weeks the stable needs to be completely mucked out, which is extremely hard work.
2. The bed can become soggy and smell if not looked after.

4.6 Two advantages of mucking out completely each day:
1. The stable floor can be kept much cleaner.
2. There is no major mucking out needed every few weeks.

Two disadvantages:
1. It takes longer on a daily basis to muck out fully.
2. More clean bedding is needed each day.

5.1 a. The procedure for measuring a horse correctly involves standing the horse on a level surface and using a measuring stick. The measuring stick must have a spirit level on the cross bar. The horse must stand squarely with his head lowered so the eyebrows are in line with the withers. The cross bar of the stick is held across the highest point of the withers. If the horse is tense he should be allowed to relax before the final measurement is recorded.
b. Three reasons why it is important horses are measured correctly:
1. It helps when selecting the correct size tack and equipment.
2. The correct measurement must be given when selling a horse.
3. Horses and ponies are often divided and subdivided in competitions according to height.

5.2 A bright bay horse is one whose body is chestnut, with black points. A dark bay has a dark brown body and black points.

5.3 A wall eye is a blue eye.

5.4 Whorls are markings made when the hair changes direction, used as identification marks.

5.5 An incorrectly tied haynet may hang too low and become entangled with the horse's feet or come undone and end up in the bedding where the horse can get entangled.

6.1 The risk of fire is high in a stable yard because many stables are of wooden construction. There is also normally a large quantity of hay and straw stored in and around the yard which burns easily and quickly.

6.2 The main danger of moving a casualty unable to get up unaided is the possibility of spinal damage in the fall. Any movement could worsen the damage and cause permanent disability.

6.3 In the event of a rider falling off your immediate course of action should include:
1. Halt the rest of the ride.
2. Assess the situation for dangers to yourself – you mustn't end up as another casualty. Move the casualty if necessary – always avoid moving if possible.
3. Assess for consciousness.
4. Check that the airway is clear.
5. Check that the casualty is breathing. If not, call an ambulance and begin CPR.
6. If the casualty is conscious find out if they have any pain and try to make them comfortable.
7. If necessary, send someone to call an ambulance.

6.4 To stem bleeding:
- Pressure should be applied to a wound to stem bleeding.
- If there is a foreign body in the wound don't try to remove it as it can worsen the damage done to arteries and veins.
- The limb should be raised if there is no underlying fracture.
- Once the bleeding has slowed, a clean dressing should be bandaged firmly over the wound.
- The casualty should then be taken to the local Accident and Emergency Department.

6.5 To make you and your horse more visible when out hacking the rider can wear fluorescent/reflective tabards, jackets and hat cover; the horse can wear fluorescent/reflective saddlecloth/exercise sheet, leg and tail bands.

6.6 Four types of weather conditions in which it is not safe to hack out on the roads:
1. Dusk or dark conditions.
2. Icy conditions.
3. Gale force winds.
4. Fog and mist.

6.7 When signalling and turning on the road the 'life-saver' look is the final look you must always give before actually turning. You are checking for traffic that you may not have seen when you first looked.

7.1 First thing in the morning the horse should appear interested in what is going on, looking forward to his breakfast or to being turned out.

7.2 Two signs indicating that there may be a problem with the horse:
1. Bedding very churned up.
2. Horse sweating.
There are other indicators including further signs of ill health.

7.3 The skin of a healthy horse should be supple and loose. Small 'ripples' should appear as you run your hand over the skin. The coat should be glossy and smooth.

7.4 The mucous membranes should appear salmon pink and moist.

7.5 The droppings of a stabled horse should be yellow-brown, fairly firm, and break on hitting the ground. Time at grass will make the droppings darker and looser.

7.6 Normal resting rates:
a. Temperature: 100.5 °F (38 °C).
b. Pulse: 25–42 beats per minute.
c. Respiration: 8–16 breaths per minute.

7.7 Two reasons why a horse may not eat his feed:
1. Ill health.
2. Dislikes feed or additives.

7.8 If a horse looks dull compared to normal you should look out for other signs of ill health. Always consult an experienced person for advice.

7.9 You should keep records of the horse's normal TPR rates as it will help you identify when there is a problem.

7.10 Two signs that could indicate the start of a bacterial or viral infection:
1. Loss of appetite.
2. Dull, disinterested attitude.

7.11 Five signs of colic:
1. Loss of appetite.
2. Dull attitude.
3. Looking round at flanks repeatedly.
4. Sweating.
5. Rolling.

7.12 The urine of a healthy horse should be more or less clear and free from odour.

7.13 Very loose droppings may be caused by worm infestation, excitement, too much rich grass, sharp teeth preventing the horse from chewing his food properly.

7.14 Four causes of poor condition:
1. Poor diet.
2. Sharp teeth, preventing the horse from chewing properly.
3. Worm infestation.
4. Overwork.

7.15 Laminitis causes the horse to move with a pottery, stilted action.

7.16 The vet should be called for:
1. Colic.
2. Deep wounds.
3. Wounds on a joint.
4. Punctured sole.
5. Infected wound.
6. Profuse bleeding.
7. If a horse is lame and you can't determine the cause.
8. Suspected fracture.
9. Repeated coughing.
10. Abnormal temperature.
11. Suspected laminitis.
12. Other signs of ill health for which you can't determine the reason or are unable to administer the necessary treatment.

7.17 Four examples of ways in which horse can demonstrate the herd instinct:
1. The horse may show signs of stress when separated from his group.
2. Horses graze and move around the paddock in a group.
3. A young horse may be reluctant to leave his group when ridden.
4. A loose horse normally heads back to his stable, to other horses or to his lorry.

7.18 a. Adrenalin and noradrenalin are known as the 'fright, flight, fight hormones'.
b. To defend himself a horse can strike out with his hooves, kick, bite, buck and rear. A horse may also try to scrape or squash a predator under or against a solid object.

7.19 a. A frightened horse may sweat; he will widen his nostrils and snort, become 'wide-eyed' and refuse to go near the object of his fear. Very often the frightened horse will try to take flight.
b. The excited horse will prance about, whinnying and pricking his ears to and fro. The horse may stale or defecate in excitement.

7.20 Two reasons a horse may show signs of tension when being ridden:
1. Pain. He may have teeth or back problems. His tack may not be fitted correctly.
2. He may lack confidence in what he is being asked to do.

7.21 a. Biting can be caused by:
- Rough and insensitive handling and grooming, in particular when tightening the girth and rugging up.
- Feeding of titbits.
- Mismanagement of young horses.

b. To control biting:
- Reprimand the horse when he bites.
- Do not feed titbits.
- Tie the horse up when grooming, etc.
- Put a muzzle on him.
- Keep a metal grille over the top of the stable door.
- Warn others that he bites.

7.22 Three causes of bucking, and their remedies:

1. Too much energy food and insufficient work. To remedy, feed non-heating, low carbohydrate food and reduce the quantities. Increase the amount of hay fed. Increase the amount of work and make the horse work harder.

2. Lack of schooling and immaturity. To remedy – work the horse on the lunge to improve schooling and help him learn to accept the saddle. When ridden, work him on a contact and keep him going positively forwards to keep his mind on his work and less inclined to buck. Check the diet as well – the young horse needs a well- balanced diet but a non-heating coarse mix or cube can be used.

3. Pain and discomfort – the saddle may be pinching or pressing or the horse may have a back problem. Check the saddle fitting and have the horse's back checked by the vet or equine chiropractor/osteopath.

8.1 Two examples of accidents that can happen whilst the horse is turned out in the field:

1. Putting a foot through the fence or gate. This is dangerous if wire fencing is used and is most likely to happen when horses are turned out in adjacent fields as they tend to fight or play over the fence.
2. Kick injuries can occur when horses don't get on together.

8.2 Three safety measures that can be taken to prevent accidents in the field:

1. Never use barbed wire as fencing.
2. Keep all fencing in a good state of repair.
3. Only turn horses out in small, amenable groups.

8.3 Barbed wire is not suitable for fencing for horses as it causes serious injuries.

8.4 Two uses of electric fencing:

1. To divide fields for resting and rotation of grazing.
2. To keep horses in adjacent fields away from the boundary fences to prevent fighting through these fences.

8.5 Shelter is needed as protection against wind, heat, flies and driving rain.

8.6 Other than a wooden field shelter, two forms of shelter:

1. Thick hedging.
2. Single screen built into the fence line.

8.7 A wooden field shelter must have a wide front opening to prevent a horse from being trapped inside by a more dominant horse.

8.8 Four poisonous plants would include any of the following: Ragwort, laurel, privet, yew, foxglove, rhododendron, laburnum. (There are also others.)

8.9 The term 'horse-sick pasture' describes a field which is over-grazed, covered in droppings and patches of long grass.

8.10 It is important that droppings are collected regularly as internal parasites (worms) continue their life-cycle through grazing horses. Larvae are passed out in the droppings of a grazing horse and are ingested by the same, or another horse. Collection of droppings drastically improves the cleanliness of a paddock and reduces worm burdens. Also, horses will not graze around their own droppings so the grass becomes long and untidy. This is also wasteful.

8.11 a. In a horse-sick paddock a lawn is an area free of droppings that has been grazed down very short.
b. A rough is an area of long grass where the horse defecates. Horses do not graze the roughs.

8.12 Each day you should check the field for:

- Security of fencing, broken rails, protruding nails.
- Water supply.
- Poisonous plants.
- Low branches.
- Rabbit holes.
- Litter.

These checks are important to keep the horse healthy, safe and to prevent accidents.

8.13 Three important points to remember when turning horses out into the paddock:

1. Lead the horse a short distance into the paddock and turn to face the fence.
2. Ensure that the gate is shut.
3. Release horses together and step backwards as you do so to avoid being kicked.

8.14 Three important points to remember when catching horses out into the paddock:

1. Do not take a bucket of feed into a group of horses as this will cause a fight and you could get kick or trampled.
2. Approach the horse from the front so he sees you.
3. If you have no option but to approach the horse from behind, speak to him so he knows you are there to avoid startling him.

9.1 The three main cereals fed to horses:

1. Oats
2. Barley
3. Maize

9.2 Bran contains high levels of phosphorus.

9.3 Sugar beet should be soaked until it has swollen fully. This normally takes 12 hours for shreds; 24 hours for cubes.

9.4 The advantages of feeding chaff:

- It aids digestion.
- It provides bulk/roughage in the diet.

- It helps to satisfy the horse's appetite without having a heating effect.

9.5 The main differences between meadow and seed hay are :
- Meadow hay is taken from pasture normally used for grazing and consists of a mixture of grasses.
- Seed hay is a specially grown crop usually consisting mainly of rye grasses.
- Meadow hay is generally softer and greener than seed hay.

9.6 Three qualities of good hay:
1. It should smell sweet, not musty.
2. The hay should fall loosely apart.
3. It must be free of poisonous weeds.

9.7 Three reasons why horses need plenty of hay in their diet:
1. Their digestive tract is designed to cope with large quantities of roughage. It therefore helps to keep the system functioning properly.
2. Eating hay helps to keep the horse occupied, mimicking the way he would act in the wild – eating more or less continuously.
3. The process of digesting hay generates heat within the body, so helping to keep the horse warm. This is important for the grass-kept horse in winter.

9.8 Changes should be introduced gradually because a sudden change would affect the horse's ability to digest the new feedstuff, which could lead to colic.

9.9 The term 'maintenance rations' refers to the ration needed purely to keep a horse healthy. The ration does not allow for extra energy for work.

9.10 Fibre is needed for healthy gut function and to provide a certain amount of energy.

9.11 Carbohydrate provides energy.

9.12 Chaff and sugar beet pulp are bulk feeds.

9.13 Oats are rolled to aid their digestion.

10.1 A mounting block should always be used when mounting as it saves the rider and horse unnecessary strain. Riding school horses are prone to back problems which can be significantly reduced by the use of the mounting block. It also lessens the chance of twisting the saddle.

12.1 Natural aids are used for communication between the rider and horse as follows:
- Seat influences – the seat can help to initiate and maintain impulsion.
- Leg influences – the legs produce impulsion by activating the horse's hindquarters. The legs also help create and maintain correct bend.
- Back influences – the back can be used to encourage the horse forwards or to slow him.
- Hand influences – the hands channel the impulsion produced by the legs and seat. The hands also give directional aids.
- Voice influences – the voice can be used to calm, reward and correct – it is essential in the initial training of the young horse.

12.2 a. The inside leg on the girth generates impulsion, helps to create the bend but also acts to stop the horse from falling in.
b. The outside leg is drawn back a little behind the girth to control the swing of the horse's quarters.
c. The inside rein asks for slight flexion in the direction of movement.
d Contact on the outside rein prevents the neck from bending too much to the inside and therefore controls the shoulder. The outside rein also helps control the speed.

12.3 a. 'Working on two tracks' means that the horse's right hooves are on one track, his left hooves on another.
b. A horse is said to be 'on the aids' when he is listening to the rider, responding to the aids given by the rider. He should be alert, yet calm and trusting – ready and able to carry out the rider's next request with ease and enthusiasm.

12.4 The reasons for selecting the outside diagonal is to ensure that the rider's seat is in the saddle when the inside hind leg is on the ground, prior to propelling the horse forward. The rider's back, seat and legs can then be used more efficiently to promote balance and impulsion.

12.5 The aids to canter:
For the novice rider this is most easily achieved if the canter strike-off is asked for in a corner of the school when the horse should be naturally bent to the inside. The horse should be established in an active working trot. The rider ceases rising to the trot and gives a preparatory half-halt before asking for the actual transition. Pressure is applied with the inside leg which remains at the girth, and the outside leg, which is drawn back slightly, behind the girth.